Analog Circuits and Signal Processing

Series Editors:

Mohammed Ismail, Dublin, USA
Mohamad Sawan, Montreal, Canada

The Analog Circuits and Signal Processing book series, formerly known as the Kluwer International Series in Engineering and Computer Science, is a high level academic and professional series publishing research on the design and applications of analog integrated circuits and signal processing circuits and systems. Typically per year we publish between 5–15 research monographs, professional books, handbooks, edited volumes and textbooks with worldwide distribution to engineers, researchers, educators, and libraries.

The book series promotes and expedites the dissemination of new research results and tutorial views in the analog field. There is an exciting and large volume of research activity in the field worldwide. Researchers are striving to bridge the gap between classical analog work and recent advances in very large scale integration (VLSI) technologies with improved analog capabilities. Analog VLSI has been recognized as a major technology for future information processing. Analog work is showing signs of dramatic changes with emphasis on interdisciplinary research efforts combining device/circuit/technology issues. Consequently, new design concepts, strategies and design tools are being unveiled.

Topics of interest include:

Analog Interface Circuits and Systems;

Data converters;

Active-RC, switched-capacitor and continuous-time integrated filters;

Mixed analog/digital VLSI;

Simulation and modeling, mixed-mode simulation;

Analog nonlinear and computational circuits and signal processing;

Analog Artificial Neural Networks/Artificial Intelligence;

Current-mode Signal Processing;

Computer-Aided Design (CAD) tools;

Analog Design in emerging technologies (Scalable CMOS, BiCMOS, GaAs, heterojunction and floating gate technologies, etc.);

Analog Design for Test;

Integrated sensors and actuators;

Analog Design Automation/Knowledge-based Systems;

Analog VLSI cell libraries;

Analog product development;

RF Front ends, Wireless communications and Microwave Circuits;

Analog behavioral modeling, Analog HDL.

More information about this series at http://www.springer.com/series/7381

Dima Kilani • Baker Mohammad
Mohammad Alhawari • Hani Saleh
Mohammed Ismail

Power Management for Wearable Electronic Devices

Springer

Dima Kilani
Khalifa University
Abu Dhabi, United Arab Emirates

Mohammad Alhawari
Wayne State University
Detroit, MI, USA

Mohammed Ismail
Wayne State University
Detroit, MI, USA

Baker Mohammad
Khalifa University
Abu Dhabi, United Arab Emirates

Hani Saleh
Khalifa University
Abu Dhabi, United Arab Emirates

ISSN 1872-082X   ISSN 2197-1854   (electronic)
Analog Circuits and Signal Processing
ISBN 978-3-030-37886-8   ISBN 978-3-030-37884-4   (eBook)
https://doi.org/10.1007/978-3-030-37884-4

This Springer imprint is published by the registered company Springer Nature Switzerland AG.
The registered company address is: Gewerbestrasse 11, 6330 Cham, Switzerland

The authors would like to dedicate this work to their parents, families, and beloved ones

Preface

With the dramatic rise of mobile electronic devices usage especially as an effect of the internet-of-things revolution, the demand for energy efficient and small form factor systems raises the need for low power multisource management unit (PMU) for energy strained devices as well as energy harvesting as alternative power source in many usage scenarios. Energy harvesting becomes one of the pillars that fulfill the needs of ultra-low power devices in many applications including the IoT-based healthcare. Since the harvested energy depends on the availability of various sources from the surroundings, a power management unit (PMU) is required to efficiently regulate the harvested energy.

Power converters and voltage regulators are important building blocks in the PMU in order to interface between the energy harvesting and the system on chip (SoC). Different types of energy harvesting source require different power converters. This depends on the electrical signal obtained from the harvester, harvester size, and efficiency. In addition, the selection of the voltage regulator depends on the area of the whole device and the requirements of various blocks in the SoC such as memory, hardware accelerator, analog front-end, and RF. Hence, sophisticated PMU circuits and techniques are required to enable the development of the state-of-the-art energy harvesting-based PMU including power converters and voltage regulators.

To accomplish this need, this book provides a comprehensive power management circuit design that targets low power wearable electronic devices powered by a thermoelectric generator (TEG) source, a battery or both. This includes extensive literature review about power converters and voltage regulators in addition to experimental results from silicon. This book is organized into 6 chapters. Each chapter carries a brief introduction of the work undertaken and is followed by the detailed circuit, results, and analysis.

Chapter 1 provides detailed background about the power management techniques at technology, circuit, and system level and delivers an overview of the recent energy harvesting source utilized for wearable electronic devices.

Chapter 2 discusses the basic concept of the TEG device and model and how it can harvest the thermal energy based on the Seebeck effect. Further, it provides a comprehensive literature review about the interface circuits required by the TEG-based PMU such as power converters, startup circuits, voltage regulators, and maximum power point tracking technique. In addition, it presents the state-of-the-art TEG-based PMU designs that are available in the literature.

Chapter 3 focuses on the characterization of the system level TEG-based PMU using several design options of power converters and voltage regulators. The characterization in terms of power efficiency, voltage ripple, and area are based on measurement results in 65 nm CMOS technology which guides the researchers to select the proper PMU design based on the blocks' requirements within the device.

Chapter 4 highlights the state-of-the-art multi-outputs switched capacitor voltage regulators. Then, it discusses a dual-outputs switched capacitor (DOSC) voltage regulator using a single switched capacitor design in order to minimize the area of PMU. It highlights how the control circuit of adaptive time multiplexing can be used to generate two output voltage levels and eliminate the reverse current problem. Measurement results are shown in 65nm CMOS technology.

Chapter 5 provides a detailed literature review on the available digital low drop out (LDO) regulator. Then, it introduces a clock-less digital LDO regulator based on a ratioed logic comparator (RLC). Simulation results are shown in 22nm FDSOI technology and a comparison with prior work on digital LDO is illustrated.

Finally, Chap. 6 concludes this book and presents possible directions for future work in this area of research.

Abu Dhabi, United Arab Emirates   Dima Kilani

Acknowledgments

The work in this book has its roots in the doctoral dissertation of the first author. We would like to thank and acknowledge all those who assisted us with the different phases of developing the material that led to this book. We would like to specifically acknowledge all members of our system on chip center (SoCC) for their help, encouragement, and support, Dr. Mihai Sanduleanu, Dr. Temesghen Habte, Ahmed Ali, Nora Harb, Huda Albanna, Yonatan kifle, and Dan Cracan. The authors acknowledge the access to the SoCC facilities for conducting the experimental work and testing the chips. Finally, we would like to acknowledge the help and support of our families and friends and thank them for their patience and understanding.

Abu Dhabi, United Arab Emirates	Dima Kilani
Abu Dhabi, United Arab Emirates	Baker Mohammad
Detroit, MI, USA	Mohammad Alhawari
Abu Dhabi, United Arab Emirates	Hani Saleh
Detroit, MI, USA	Mohammed Ismail

Contents

Abbreviations

AFE	Analog Front-End
ATM	Adaptive Time Multiplexing
CMOS	Complementary Metal Oxide Semiconductor
DLDO	Digital Low Drop Out
DVS	Dynamic Voltage Scaling
LDO	Low Drop Out
PEG	Piezo Electric Generator
PFM	Pulse Frequency Modulation
PMU	Power Management Unit
RLC	Ratioed Logic Comparator
SC	Switched Capacitor
SI	Switched Inductor
SoC	System on Chip
TEG	Thermo Electric Generator

Symbols

V_{dd}	Supply voltage	V
V_{ds}	Drain–source voltage	V
V_{gs}	Gate–source voltage	V
V_{th}	Threshold voltage	V
V_{in}	Input voltage	V
V_L	Load voltage	V
V_o	Output voltage	V
V_{ref}	Reference voltage	V
V_T	Thermal voltage	V
V_{TEG}	TEG voltage voltage	V
V_{boost}	Boosted voltage	V
$V_{ratioed\text{-}logic}$	Output voltage of ratioed logic circuit	V
I_{in}	Input current	A
I_{out}	Output current	A
I_q	Quiescent current	A
I_{ds}	Drain–source current	A
$P_{average}$	Average power consumption	W
P_{active}	Power consumption during active mode	W
P_{sleep}	Power consumption during sleep mode	W
$P_{transition}$	Power consumption from active	W
	mode to sleep mode and vice verse	W
$P_{always\text{-}on}$	Power consumption of the always-on blocks	W
P_{in}	Input power	W
P_{out}	Output power	W
$P_{dynamic}$	Dynamic power	W
$P_{short\text{-}circuit}$	Short circuit power	W

P_{cfly}	Flying capacitor power losses	W
P_{cond}	Conduction power losses	W
P_{sw}	Switching power losses	W
P_{bp}	Bottom palate capacitor power losses	W
C_{fly}	Flying capacitor	F
C_L	Load capacitor	F
C_{ox}	Oxide-gate capacitor	F
L	Inductor	H
R_{ds}	Drain–source resistance	Ω
R_{TEG}	TEG resistance	Ω
f	Switching frequency	Hz
W	Width of the transistor	m
L	Length of the transistor	m
μ	Electrical mobility	$\frac{m^2}{V \cdot s}$
η	Efficiency	
Φ_1	Gate-drive pulse of switch (charging phase)	
Φ_2	Gate-drive pulse of switch (discharging phase)	

List of Figures

List of Tables

Chapter 1
Introduction to Power Management

The advancements in CMOS technology have enabled the evolution of low power circuits and systems in many applications including wearable biomedical devices. It allows the devices to operate at sub-microwatt power range and opens up a new era of the Internet of things. These new devices require small size with near perpetual lifetime. However, the traditional battery technology is not scaling at the same rapid rate as the shrinking of transistor size. Therefore, there is a need for efficient management of the power source to reduce both dynamic and leakage power of the device and; a need to increase the energy sources via energy harvesting. The introduction of energy harvesting brings some design challenges since the harvested energy has variable voltage and it depends on the availability and size of the harvester. Therefore, a power management unit is needed to manage the energy transfer from the harvester and control the power distribution and consumption of different blocks in the device.

In this chapter, we introduce the wearable electronic devices' trends and constraints. Then, we explore the need of a power management unit for a low power wearable electronic device to achieve high power efficiency and small form factor. After that, the aim and objectives of this book are discussed.

1.1 Low Power Wearable Electronic Devices

We are living in the era where electronic devices are an integrated part of all aspects of our lives. It makes our daily lives much easier when we travel, study, work, exercise, communicate, and monitor or treat illnesses. Wearable electronics such as wristwatches, necklaces, and earbuds are widely utilized in healthcare applications. These devices have a good potential in monitoring the patient's health, allowing personal diagnosis, and tracking user's fitness and exercises. They also enable a wireless interface between the patients/doctors and device for remote monitoring

© Springer Nature Switzerland AG 2020   1
D. Kilani et al., *Power Management for Wearable Electronic Devices*, Analog Circuits
and Signal Processing, https://doi.org/10.1007/978-3-030-37884-4_1

and control. This helps the doctors to follow up on the patient's health status without the need for the patient to physically visit the hospital.

Although the projection of wearable devices is expected to increase in the future, some design challenges exist. First, the wearable electronic devices are typically powered by a battery. Despite the improvement in the battery capacity in the last decade [1], it runs out of energy after a fairly limited certain period of time and operation. This limits the lifespan of the wearable devices as they require continuous monitoring and sensing. As a result, the battery needs to be replaced or converted to a rechargeable battery which increases maintenance and cost. Therefore, battery lifetime is a critical issue in wearable devices that target long life or near perpetual operation. Second, the battery solution significantly increases the size and cost of the devices which makes it inconvenient for the human to wear it. Hence, the small form factor is another critical issue for wearable devices. The main constraints that will be focused in this book are the power consumption and the form factor of the wearable electronics as part of the power management unit (PMU).

Figure 1.1 shows the general block diagram of a wearable device. It consists of an energy source and power management, sensors, an analog front end (AFE), a digital processor, and a wireless transceiver. The energy source powers up the whole device and the power management distributes this power to different blocks in the system in an efficient way according to their required performance. The sensors collect data and send it to the digital processor where the digitized data are processed and saved in the memory if needed. Finally, the wireless transceiver sends the processing results to the user/receiver. Given the block diagram in Fig. 1.1, Fig. 1.2 shows the current consumption breakdown of different blocks in a wearable device as reported in [2]. As shown in the figure, most of the blocks consume low current in the range of μA. The low power transmitter TX consumes $160\,\mu$W at a maximum data rate of 200 kbps whereas it consumes 190 nW when the device is duty cycled by

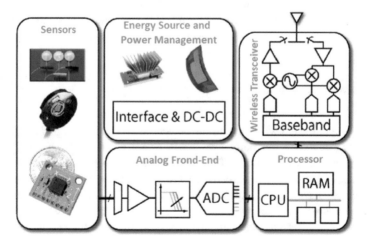

Fig. 1.1 General block diagram of a wearable device [5]

Fig. 1.2 Current consumption breakdown of different blocks in a low power wearable device [2]

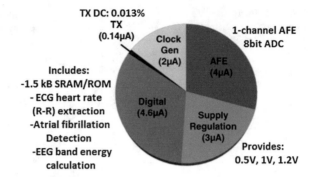

0.013% [3]. Usually, the wireless transmitter consumes a large amount of power and therefore it operates in a duty cycle fashion in order to minimize the overall power consumption of the device [4].

1.2   Low Power Management Techniques

Several approaches are utilized to reduce the battery's size and increase the battery's lifetime. One of these approaches is to reduce the overall power consumption of the integrated circuit and system in the wearable device. This way, the current derived from the battery is reduced and the battery's lifetime is extended. Therefore, a PMU is needed to control and manage the power consumption of different blocks in the device in order to maximize power efficiency.

1.2.1   Dynamic Power Reduction

There are two main power dissipation components in an integrated circuit: dynamic power and leakage power. The dynamic power is defined as the power consumed by the circuit when it is switching. Dynamic power includes two components which are switching power P_{sw} and short circuit power $P_{short-circuit}$ as given in Eq. (1.1). Switching power that is given in Eq. (1.2) is related to charging and discharging the circuit where α is the activity factor of the circuit, C is the switching capacitance, V_{dd} is the supply voltage, and f is the switching frequency. The short circuit power is related to the current that flows directly from the supply voltage to the ground when both NMOS and PMOS networks are momentarily on during the switching. The loss of short circuit power is relatively small and the switching power dominates the dynamic power consumption.

$$P_{dynamic} = P_{sw} + P_{short-circuit} \qquad (1.1)$$

Fig. 1.3 Energy per
operation versus supply
voltage

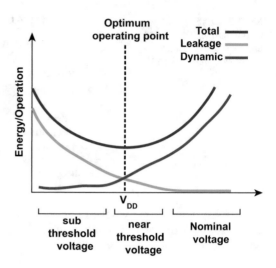

$$P_{sw} = \alpha C V_{dd}^2 f \tag{1.2}$$

The switching power is scaled in a quadratic manner with the supply voltage but at
the cost of speed. Frequency scaling can be used in conjunction with voltage scaling
to further reduce power consumption. Reducing voltage and frequency is referred
to dynamic voltage-frequency scaling (DVFS). The advantage of DVFS is that
dynamic power can be significantly reduced according to the device performance
requirement. In other words, DVFS allows the device to run at an operating
point while the required performance is achieved as shown in Fig. 1.3. However,
increasing voltage scaling and operating at subthreshold voltage increases the
leakage energy since the device operates longer and slower at this region. Therefore,
it is more efficient to operate near the subthreshold region where both leakage and
dynamic power are optimum.

For a wearable device that is a highly power-constrained device, supply voltage
scaling is needed to provide the desired voltage level for each block rather than
powering the whole device with one voltage level. The SRAM operates at a
minimum voltage level of 0.9 V in 45 nm CMOS [6] whereas the digital blocks can
operate with as low as 0.4 V [7] to minimize power consumption. Since the wireless
transceiver consumes a significant amount of power, it requires a high voltage level
such as 1 V [4]. Therefore, having multiple voltage levels in a single chip is required.
Figure 1.4 depicts the power architecture of a 16-cores ARM microprocessor where
each core requires a DVFS block. Dynamic voltage scaling (DVS) can be achieved
using a voltage regulator that has the ability to scale the input voltage into a lower
output voltage level. The integration of multiple voltage regulators is needed to
support DVS per block. Thus, the area of PMU is dominated by the area of voltage
regulators. As the number of the desired regulated voltage levels increases, the
number of voltage regulators increases which results in a large area overhead.

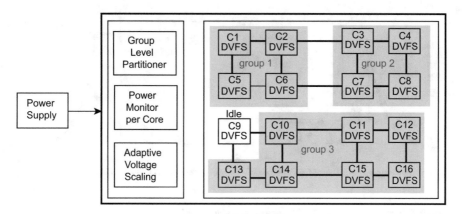

Fig. 1.4 Power architecture for 16-chip core ARM multiprocessor [8]

Therefore, supporting multiple regulated voltage levels with a minimum number of voltage regulators needs to be explored.

1.2.2   Leakage Power Reduction

Leakage power is defined as the power consumed by the circuit when it is inactive. Leakage power can be also found when the circuit is operating but part of it is inactive. This type of power is called active leakage. Generally, due to the large integration of transistors in a single chip, leakage power becomes critical especially in multi-cores systems-on-chip (SoC) as some cores remain inactive which results in a dark silicon issue [9]. In addition, although the continuous advancements in CMOS technology in shrinking the transistor's size improve the device speed, the leakage current is increasing rapidly. This is because the channel length is decreasing and allowing more leakage current from the source to drain or from the source to body. The impact of leakage power becomes even worse in low power devices as the average power consumption is dominated by the power during idle time. Therefore, a duty cycling technique is employed to ensure that the system enters a low power mode during the idle time where the supply voltage is reduced for some blocks and disconnected for unnecessary blocks. Figure 1.5 shows the activity of a low power wearable device that relies on a low duty cycling technique. As depicted in the figure, the device wakes-up momentarily for a few milliseconds to perform a repetitive task of sensing, processing, and communicating and then, falls asleep till a next operational task. The average power consumption of the device $P_{average}$ is given in Eq. (1.3) where P_{active} is the power consumption of the device during active mode, P_{sleep} is the power consumption of the device during sleep mode, $P_{transition}$ is the power consumption of the device when it transits from active to sleep mode and vice versa, and $P_{always-on}$ is the power consumption of the blocks

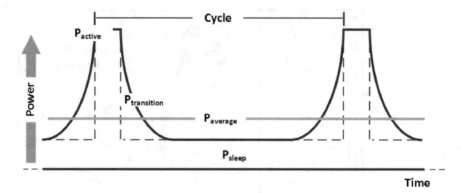

Fig. 1.5 Typical duty cycle of a low power wearable device

that are always on such as memory and voltage regulators. Note that the average power consumption of the device is dominated by the sleep power due to the low duty cycle (0.1–1%) of the device [10]. As such, reducing the power consumption of the device during the sleep mode would enhance the total power efficiency.

$$P_{average} = P_{active} + P_{sleep} + P_{transition} + P_{always\text{-}on} \qquad (1.3)$$

However, the wake-up time that the device consumes from sleep to active mode is larger than the time it takes from idle to active mode. Thus, there is a tradeoff between power consumption and wake-up time. Therefore, having an idle mode in addition to sleep and active modes in some devices is required for a faster wake-up time. Figure 1.6 depicts 6 power management modes for 6-cores atom processor. Each core operates in a particular power mode with certain voltage and frequency levels. The low power mode in C6 has the longest wake-up time as most of the blocks are turned off.

Providing different power modes to minimize both active and leakage power adds more control circuits and techniques to PMU such as power gating where a header or a footer transistor is used to disconnect the supply voltage from the block. This way, the leakage power is minimized especially if a high threshold header/footer device is implemented. However, power gating degrades the performance during active mode due to the use of high threshold header/footer device. In addition, different power modes can be supported at system level power management by establishing a finite state machine (FSM). The FSM is an important block to interface between PMU and SoC. Based on the workload of each block in the SoC, digital signals are generated from the SoC and sent as *Inputs* to the PMU as shown in Fig. 1.7. The PMU determines the required operating power modes using FSM (sometimes it is called power state machine) and generates digital signals *Outputs* to the SoC to control the voltage, frequency, and functionality level of each block.

In addition to power gating and multiple power modes, raising threshold voltage V_{th} is another method to reduce leakage current and can be implemented using

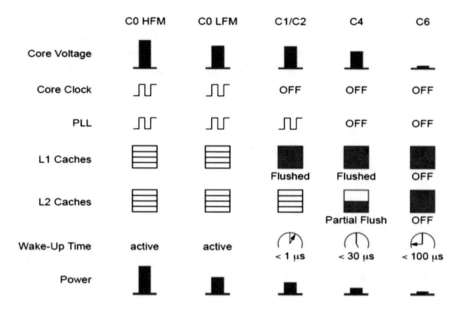

Fig. 1.6 Atom power management modes [11]

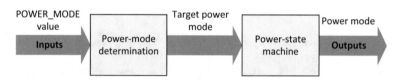

Fig. 1.7 Power management flow [12]

source biasing, stacking, and body effect. These techniques increase the threshold voltage level which reduces leakage current but at the cost of performance degradation during the active mode. Usually, each device has an acceptable percentage of performance degradation so that additional delay might have little change on the overall circuit.

1. **Source Biasing**

   Biasing the source of the transistor can significantly reduce the leakage current. When source biasing is implemented, the gate–source voltage V_{gs} is reduced which has an exponential impact on reducing the subthreshold current. Another result of source biasing is that the drain–source voltage V_{ds} is reduced which decreases the leakage current. Moreover, applying a voltage to the source reduces the body–source voltage V_{bs} resulting in a higher V_{th} due to body effect as discussed in the body effect section.

2. **Stacking**

Connecting a set of transistors in a series manner is referred to stacking. Figure 1.8 compares between the standard NMOS with 2-stacked NMOS transistors. As depicted in Fig. 1.8a, when NMOS is off, $V_{gs} = 0$. In the case of the 2-stacked configuration as shown in Fig. 1.8b, $V_{gs1} = -V_x$ instead of 0 which makes $M1$ in the super cutoff region so that further leakage reduction is achieved. In addition, the source–body voltage of $M1$, V_{sb1}, becomes less in the negative value rather 0 which also reduces leakage current. The reduction in leakage is commonly called the stack effect. However, the stacking technique introduces more delay since additional devices are added.

3. **Body Effect**

The source biasing and stacking described previously reduces simultaneously both V_{gs} and V_{sb}. However, there is another technique specialized in reducing only V_{sb} which is called body biasing. Applying voltage to the body of the transistor has a direct proportional relationship to the threshold voltage as given in Eq. (1.4) where V_{T0} is the threshold voltage for $V_{sb} = 0$, ϕ_F is the substrate Fermi potential and γ is the body effect coefficient. In the case of NMOS transistor, a negative voltage is applied to the body and for PMOS transistor, a positive voltage is applied. This way, the threshold voltage is increased and the leakage current is reduced.

$$V_{th} = V_{T0} + (\gamma * \sqrt{|2\phi_F + V_{sb}|} + \sqrt{|2\phi_F|}) \tag{1.4}$$

Apart from circuit and system level power management, the development of CMOS technology in the last five decades plays an important role in PMU design from the technology level of abstraction. The CMOS technology advancement enables low voltage operation at the near-threshold region to reduce active power. It also introduces multi-threshold devices to cut down leakage power. The introduction

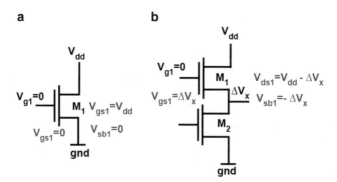

Fig. 1.8 (a) NMOS transistor; (b) two stacked NMOS transistors

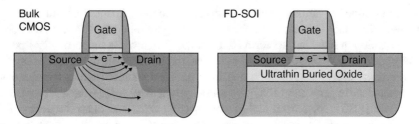

Fig. 1.9 Traditional bulk CMOS versus FDSOI [13]

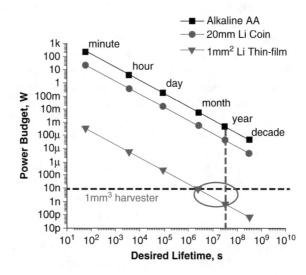

Fig. 1.10 Power budget versus lifetime for the harvester and different batteries [1]

of fully depleted silicon-on-insulator (FDSOI) technology that adds a very thin oxide layer beneath the silicon transistor reduces the leakage current compared to bulk CMOS as shown in Fig. 1.9. FDSOI enables reverse body biasing (high threshold voltage device), on the one hand, to reduce leakage power when high performance is unnecessary and enables forward body biasing (low threshold voltage), on the other hand, to maximize the performance at expense of leakage power. However, as the body bias varies as a new variable, the corners become process–voltage–temperature–body bias (PVTB) which should be taken into consideration.

1.3 Energy Harvesting-Based Device

Another method to overcome the issue of the battery's size and lifetime is the utilization of energy harvesting which recharges or replaces the battery [14–16]. The low power consumption of wearable devices enables the use of energy harvesting as the main energy source. Figure 1.10 shows the power budget versus the energy source lifetime for various types of battery: Alkaline AA, 20 mm Lithium Coin, and a 1 mm^2 Lithium Thin-film. As shown in the figure, energy harvester becomes a viable option in reducing the form factor of the device while supporting low power budget (nW to μW range) required by the wearable electronic devices [17, 18].

Electrical energy can be harvested from different sources such as thermal from a thermoelectric generator (TEG); vibration, pressure, or force from a piezoelectric generator (PEG); solar from a photovoltaic (PV) cells; and radio frequency (RF) from a wireless RF. Selecting the energy harvesting source depends on the harvested power and source availability, device power budget, and device size. Table 1.1 shows the harvested power density for different energy harvesting sources. As shown in Table 1.1, the ambient light source provides the highest harvested power density in both indoor and outdoor which can be used in application that requires mW range. The main drawback of this harvester type is that the amount of solar energy strongly depends on the light availability which is limited during dark or low lighting conditions. Therefore, other sources such as thermal or vibration can be considered. The thermal and vibration energy sources are good options for wearable devices that require a low power budget (μW as referred in Fig. 1.2) since the harvested power from the human is in the μW range. The harvested power that should be equal or greater than the device average power consumption depends on the harvester size. Thus, there is a tradeoff between the power and size of the harvester. Therefore, a PMU is needed to implement efficient interface circuits between the harvester source and the SoC to extract the maximum power while maintaining the device small form factor. Since the amount of the harvested power is variable and depends on the availability of the harvester source and the interface circuit, voltage regulators are required to support a steady and regulated voltage level to supply the SoC.

Figure 1.11 shows the general block diagram of an energy harvesting-based PMU. It consists of a power converter, a storage element, a voltage regulator, and a startup circuit. The power converter is used to convert the energy from one form to another. For example, the AC–DC converter (rectifier) is utilized in devices powered

Table 1.1 Harvested power density for different energy harvesting sources [19]

Source	Source power density	Harvested power density
Ambient light		
Indoor	0.1 mW/cm^2	10 μW/cm^2
Outdoor	100 mW/cm^2	10 mW/cm^2
Vibration		
Human	0.5 m at 1 Hz	4 μW/cm^2
Industry	1 m/s^2 at 50 Hz	100 μW/cm^2
Thermal		
Human	20 mW/cm^2	30 μW/cm^2
Industry	100 mW/cm^2	1–10 mW/cm^2
RF		
GSM base station	0.3 μW/cm^2	0.1 μW/cm^2

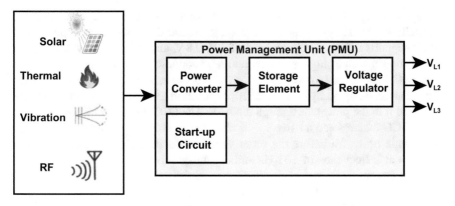

Fig. 1.11 General block diagram of an energy harvesting-based PMU

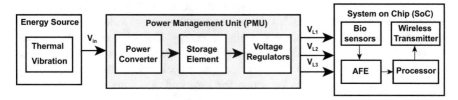

Fig. 1.12 Proposed block diagram of a self-powered wearable biomedical device that detects and predicts the heart attack of the human body

by vibration or RF. In addition, power converter implements control circuits to extract the maximum power point and stores it in a storage element either a battery or a super-capacitor. The battery can be recharged by the energy harvesting; however, it suffers from large size and results in a larger form factor which is not desirable in the wearable device. The battery can be replaced by a super-capacitor as it has smaller area and lower charging and discharging time. However, it has smaller energy density than the battery. Typically, the output from power converter is not regulated so a voltage regulator is employed to support a regulated output voltage. This means that the voltage regulators are not only used for DVS as mentioned in Sect. 1.2 but also for voltage regulation in energy harvesting-based devices. Finally, the startup circuit is utilized to start and power up the whole device at the beginning.

1.4   Aim and Objectives

Figure 1.12 shows the proposed block diagram of a low power wearable device that detects the heart attack of the human body. It consists of energy source, PMU, and SoC. The device is powered by the thermal and vibration energies obtained from the human body. The circuit efficiently transfers the harvested energy into the main storage unit. The PMU regulates the output voltages for the system and supports normally-off instantly-on operation mode to reduce the power consumption. The PMU supports different voltage level of 1 V, 0.8 V, and 0.6 V to provide different power modes such as high, medium, and retention/low power modes, respectively. The SoC includes a biomedical sensor, a processor, and a wireless transmitter. Graphene oxide nylon ECG sensors [20] are utilized to detect data from the patient and send it to the processor through the AFE. The ECG processor utilizes a unique set of ECG features and a naive Bayes classifier in order to predict the ventricular arrhythmia up to 3 h before the onset as published in [21]. The ECG processor operates at a frequency of 10 KHz and consumes a power of 2.78 μW [21]. The wireless transmitter transfers the ECG data of the patient and alerts the hospital or physician in case of any emergency. The wireless transceiver operates at a voltage level of 1 V (high power mode). The memory can run either at a voltage of 1 V (high power mode) or 0.6 V (retention mode). On the other hand, the digital processor and the sensor operate at a supply voltage of 1 V (high power mode), 0.8 V (medium power mode), or 0.6 V (low power mode) to enable voltage scaling based on system performance needs and available energy. The AFE and ADC operate at 0.6 V (low power mode). The targeted system specifications have an average power consumption of 6 μW for one complete cycle where the SoC including the transmitter is assumed to be active for 1 ms and then falls asleep except the AFE/ADC, sensor, and PMU for 2 s till the next task.

The aim of this book is to build an efficient PMU for a low power wearable biomedical device given in Fig. 1.12 that targets μW range. High power efficiency and small area are the two main factors targeted in the PMU design. The efficiency and area of PMU depend mainly on the efficiency and area of voltage regulators. Therefore, this book focuses on the circuit design of various monolithic voltage regulators that provide DVS and regulation targeting μW load power range. This book includes two main objectives. The first book objective is to design efficient voltage regulators generating multiple output voltages required by the proposed device with low power and area overhead. The second book objective is to characterize and evaluate several combinations of power converter and voltage regulators powered by thermal energy harvesting. This characterizations help to select the appropriate PMU design option according to the blocks' requirements within the device.

References

1. D. Sylvester, *How to design nanowatt microsystems*. University of Michigan, 2013.
2. Y. Zhang, F. Zhang, Y. Shakhsheer, J. D. Silver, A. Klinefelter, M. Nagaraju, J. Boley, J. Pandey, A. Shrivastava, E. J. Carlson *et al.*, "A batteryless 19 μW MICS/ISM-band energy harvesting body sensor node SoC for ExG applications," *IEEE Journal of Solid-State Circuits*, vol. 48, no. 1, pp. 199–213, 2013.
3. J. Pandey and B. P. Otis, "A sub-100 μw MICS/ISM band transmitter based on injection-locking and frequency multiplication," *IEEE Journal of Solid-State Circuits*, vol. 46, no. 5, pp. 1049–1058, 2011.

4. J. Yi, F. Su, Y.-H. Lam, W.-H. Ki, and C.-Y. Tsui, "An energy-adaptive MPPT power management unit for micro-power vibration energy harvesting," in *Circuits and Systems, 2008. ISCAS 2008. IEEE International Symposium on*. IEEE, 2008, pp. 2570–2573.
5. Y. Ramadass. (2013) Energy harvesters and energy processing circuits. [Online]. Available: http://www.isscc.com.ISSCCtutorial.132
6. B. Mohammad and J. Abraham, "A reduced voltage swing circuit using a single supply to enable lower voltage operation for SRAM-based memory," *Microelectronics Journal*, vol. 43, no. 2, pp. 110–118, 2012.
7. M. Ashouei, J. Hulzink, M. Konijnenburg, J. Zhou, F. Duarte, A. Breeschoten, J. Huisken, J. Stuyt, H. de Groot, F. Barat *et al.*, "A voltage-scalable biomedical signal processor running ECG using 13pJ/cycle at 1 MHz and 0.4 V," in *Solid-State Circuits Conference Digest of Technical Papers (ISSCC), 2011 IEEE International*. IEEE, 2011, pp. 332–334.
8. K. Ma, X. Li, M. Chen, and X. Wang, "Scalable power control for many-core architectures running multi-threaded applications," in *ACM SIGARCH Computer Architecture News*, vol. 39, no. 3. ACM, 2011, pp. 449–460.
9. A. Rahmani, P. Liljeberg, A. Hemani, A. Jantsch, and H. Tenhunen, *The Dark Side of Silicon*. Springer, 2016.
10. M. Alioto, "Ultra-low power VLSI circuit design demystified and explained: A tutorial," *IEEE Transactions on Circuits and Systems I: Regular Papers*, vol. 59, no. 1, pp. 3–29, 2012.
11. T. R. Halfhill, "New low-power microarchitecture rejuvenates the embedded x86." Intel, 2008.
12. D. Macko, K. Jelemenská, and P. Cicák, "Power-management high-level synthesis," in *Very Large Scale Integration (VLSI-SoC), 2015 IFIP/IEEE International Conference on*. IEEE, 2015, pp. 63–68.
13. L. Gwennap, "FD-SOI offers alternative to FINFET," *Posted at* https://www.globalfoundries.com/sites/default/files/fd-soi-offers-alternative-tofinfet.pdf, 2016.
14. A. Camarda, M. Tartagni, and A. Romani, "A- 8 mv/+ 15 mv double polarity piezoelectric transformer-based step-up oscillator for energy harvesting applications," *IEEE Transactions on Circuits and Systems I: Regular Papers*, vol. 65, no. 4, pp. 1454–1467, 2018.
15. J. Katic, S. Rodriguez, and A. Rusu, "A high-efficiency energy harvesting interface for implanted biofuel cell and thermal harvesters," *IEEE Transactions on Power Electronics*, vol. 33, no. 5, pp. 4125–4134, 2018.
16. D. El-Damak and A. P. Chandrakasan, "A 10 nw–1 μw power management IC with integrated battery management and self-startup for energy harvesting applications," *IEEE Journal of Solid-State Circuits*, vol. 51, no. 4, pp. 943–954, 2016.
17. A. Wang, J. Kwong, and A. Chandrakasan, "Out of thin air: Energy scavenging and the path to ultralow-voltage operation," *IEEE Solid-State Circuits Magazine*, vol. 4, no. 2, pp. 38–42, 2012.
18. A. P. Chandrakasan, N. Verma, and D. C. Daly, "Ultralow-power electronics for biomedical applications," *Annual review of biomedical engineering*, vol. 10, 2008.
19. R. J. Vullers, R. Van Schaijk, H. J. Visser, J. Penders, and C. Van Hoof, "Energy harvesting for autonomous wireless sensor networks," *IEEE Solid-State Circuits Magazine*, vol. 2, no. 2, pp. 29–38, 2010.
20. N. Hallfors, M. A. Jaoude, K. Liao, M. Ismail, and A. Isakovic, "Graphene oxide—nylon ECG sensors for wearable IoT healthcare," in *Sensors Networks Smart and Emerging Technologies (SENSET), 2017*. IEEE, 2017, pp. 1–4.
21. N. Bayasi, T. Tekeste, H. Saleh, B. Mohammad, A. Khandoker, and M. Ismail, "Low-power ECG-based processor for predicting ventricular arrhythmia," *IEEE Transactions on Very Large Scale Integration (VLSI) Systems*, vol. 24, no. 5, pp. 1962–1974, 2016.

Chapter 2
Introduction to TEG-Based Power Management Unit

This chapter discusses in details the state of the art thermoelectric generator (TEG)-based power management unit (PMU) circuits in the literature. First, we introduce the basic concept of the TEG device and how the electricity is generated from heat. Then, we explain the interface circuit needed when using TEG device. Finally, we present the existed TEG-based PMU in the literature.

2.1 Introduction to TEG

The temperature difference surrounds us everywhere in nature, environment, home, and industry. This temperature difference can be converted into electrical energy based on the Seebeck effect using TEG devices. The basic concept of Seebeck effect is that when the temperature difference is found between two dissimilar metals or semiconductors, a voltage difference is generated across these two materials. Although this concept has been well known and was first discovered by Thomas John Seebeck in 1822 [1], the advantages of the TEG show great potential in many applications especially wearable electronics as it provides a continuous source of heat from the human body. In addition, it does not require moving parts due to its solid state design and hence requires no maintenance and no extra cost [2]. Furthermore, TEG systems support perpetual and autonomous operations and it is also environmentally friendly to use as it can work in any position [2].

Figure 2.1a shows a TEG device structure consisting of a single n-type and p-type legs connected in series between two heat exchangers. When the temperature is applied to the hot side, the electrons and holes in the n-type and p-type semiconductors, respectively, get stimulated and travel to the cold side. The charge concentration at the cold side produces an electric potential V_{TEG}. This voltage is proportional to the temperature difference ΔT as shown in Eq. (2.1) where S is the Seebeck coefficient which is defined as the voltage change per temperature degree

© Springer Nature Switzerland AG 2020
D. Kilani et al., *Power Management for Wearable Electronic Devices*, Analog Circuits and Signal Processing, https://doi.org/10.1007/978-3-030-37884-4_2

Fig. 2.1 (**a**) TEG device
using single n-type and
p-type legs; (**b**) TEG Device
using multiple n-type and
p-type legs; (**c**) TEG model
using voltage source and
resistor in series

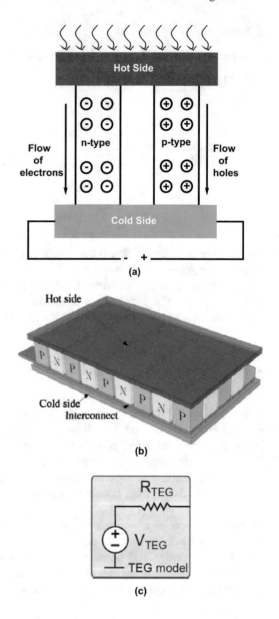

of the material. Figure 2.1b shows the typical TEG devices consisting of a set of n-type and p-type legs connected electrically in series and thermally in parallel since the heat flows from top to bottom. Finally, Fig. 2.1c shows the TEG electrical model which consists of a voltage source and a series resistance. This model will be used in this work during our TEG-based PMU characterizations in Chap. 3.

$$V_{TEG} = S \times \Delta T \tag{2.1}$$

2.2 Interface Circuit of TEG

Since V_{TEG} depends on the temperature difference as given in Eq. (2.1), it creates a major problem when using it in wearable devices. This is due to the small temperature difference obtained from the human body and results in a low V_{TEG} level <100 mV [3]. This low voltage level cannot be used directly to supply CMOS circuit. Therefore, a startup circuit is required to kick off the system and interface between the TEG and CMOS circuit. Figure 2.2 shows the interface circuit of the TEG-based PMU. It includes a startup circuit, a maximum power point tracking (MPPT), a DC–DC boost converter, and a voltage regulator. The startup circuit kicks off the boost converter. The boost converter including MPPT generates a boosted voltage V_{boost} from V_{TEG} and extract the maximum power. In addition, a voltage regulator is used to support a regulated load voltage V_L from the unregulated V_{boost}. Each block in the TEG-based PMU as shown in Fig. 2.2 is discussed in the coming sections.

2.2.1 Startup Circuit

There are many startup circuit techniques reported in the literature. The work in [4] recharges the load capacitor for one time to start up the boost converter. The work in [5] utilizes a motion activated switch to start up the wearable application from the vibration of the human body. A transformer element is used to start up the boost converter in [6]. However, it is an off-chip component which would increase the size of the wearable applications and that is not favorable in our system. The charge pump circuit is utilized in [7] and [8] to kick off the boost converter. The minimum TEG voltage required is 250 mV in [7] and 300 mV in [8] which is not suitable for our wearable application where the TEG voltage is between 50 and

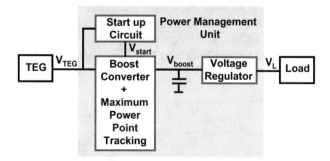

Fig. 2.2 General block diagram of a TEG-based PMU including boost converter with maximum power point tracking, startup circuit, and voltage regulator

65 mV. In addition, the work in [9] implements an LC oscillator followed by a voltage multiplier from an extremely low V_{TEG} of 50 mV to generate a high voltage. This design requires an off-chip inductor and adds more parasitics.

2.2.2 DC–DC Boost Converter

There are two types of DC–DC boost converter: charge pump and inductor-based boost converter. The advantages and disadvantages of each type are explained in details in the coming sections.

2.2.2.1 Charge Pump

The charge pump utilizes diodes (or diode-connected transistors) and capacitors to transfer the charges through charging and discharging phases. Figure 2.3 shows a Dickson charge pump voltage doubler topology. When $\phi_1 = 0$ and $\phi_2 = 1$, D1 is on where C_1 is charged to V_{in} assuming ideal diodes. During this phase, D_2 is off. When $\phi_1 = 1$ and $\phi_2 = 0$, C_1 becomes in series with V_{in} and D_2 is on where C_2 is charged to $2V_{in}$. Thus, the output voltage is $2V_{in}$. The output voltage is proportional to the number of stages N in the pump but it is reduced by NV_t assuming non-ideal diode, where V_t is the voltage drop across the diode. The output voltage of the charge pump at no output current is given in Eq. (2.2).

$$V_{out} = (N + 1)(V_{in} - V_t) \tag{2.2}$$

The advantage of the charge pump is that it can provide an integrated solution since the capacitors can be fully integrated on-chip. However, it is inefficient due to the inherent energy losses when the charges are transferred from one capacitor to another. These inherent energy losses exist even when all components of the circuit are ideal [10]. In addition, the charge pump has a limited driving current capability since it depends on the capacitance density of the on-chip capacitors. Further, according to Eq. (2.2), a 19 stages charge pump is required in our wearable device to boost the TEG voltage from 50 mV to 1 V which costs larger area.

Fig. 2.3 Charge pump circuit

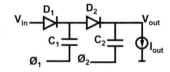

2.2.2.2 Inductor-Based Boost Converter

Conversely, inductor-based boost converter utilizes switches and inductor to store the energy and transfer it to the load. Theoretically, it achieves 100% power efficiency. In addition, it provides high boost ratio and therefore, it is widely utilized in TEG-based applications [11]. The efficiency of the inductor-based boost converter depends on the quality factor of the inductor, control power losses, and technology used [12]. However, the inductor cannot be integrated on-chip as the design requires an inductor's value in the μH range. Therefore, there is a tradeoff between integration and power efficiency.

There are two types of inductor-based boost converter: asynchronous and synchronous boost converters. The asynchronous boost converter is shown in Fig. 2.4a. It consists of an inductor, a switch M_1, a diode, and a load capacitor. When clk_1 is high as shown in Fig. 2.4b, M_1 is turned on and the current flows from the voltage source to the inductor where the energy is stored. Note that the diode, in this case, is off and disallows the current to pass to the load. When clk_1 is off as shown in Fig. 2.4c, M_1 is off and diode becomes on so that the inductor discharges the current into the load. Figure 2.4d shows the waveform of the inductor current I_{in} and the

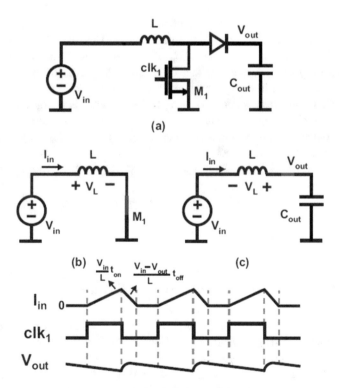

Fig. 2.4 (**a**) Asynchronous inductor-based boost converter; (**b**) charging phase; (**c**) discharging phase; (**d**) inductor current and output voltage waveform [13]

Fig. 2.5 (**a**) Synchronous inductor-based boost converter; (**b**) inductor current and output voltage waveform [13]

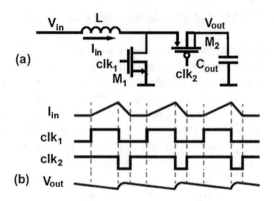

output voltage V_{out}. This implies that the output voltage depends on the duty cycle D of M_1 as given in Eq. (2.3). The duty cycle is defined as the ratio of T_{on} to the clock period T.

$$V_{out} = \frac{V_{in}}{1 - D} \qquad (2.3)$$

The advantage of the asynchronous boost converter is that one clock signal is required which makes the design easy to implement. However, the diode degrades power efficiency due to its voltage drop and internal resistance. To improve the power efficiency and reduce the voltage drop across the diode, the diode is replaced by a PMOS transistor in the synchronous inductor-based boost converter as shown in Fig. 2.5a. In this case, when clk_1 is high, the inductor is charged through M_1, whereas when clk_2 is low, the inductor releases its energy to the load. clk_1 and clk_2 are synchronized at the falling edge where clk_2 falls once clk_1 falls. The output voltage of the synchronous boost converter depends on the on-time of the NMOS and PMOS transistors $T_{nmos\text{-}on}$ and $T_{pmos\text{-}on}$, respectively, as given in Eq. (2.4) [4].

$$V_{out} = (1 + \frac{T_{nmos\text{-}on}}{T_{pmos\text{-}on}}) V_{in} \qquad (2.4)$$

2.2.3 Maximum Power Point Transfer Techniques

In order to extract the maximum power point, the impedance of the TEG device R_{TEG} should match the input impedance of the boost converter R_{in}. Both impedances are matched ($R_{TEG} = R_{in}$) at $V_{in} = \frac{V_{TEG}}{2}$ as shown in Fig. 2.6, where V_{in} is the input voltage of the boost converter. At this point, the maximum power from the TEG is achieved as given in Eq. (2.5).

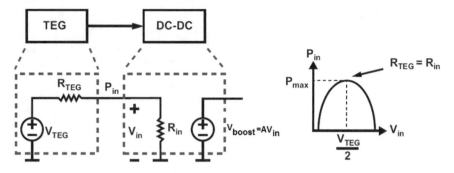

Fig. 2.6 Models of TEG and DC–DC converter and the maximum power extraction graph [13]

$$P_{in\text{-}max} = \frac{V_{TEG}^2}{4R_{TEG}} \qquad (2.5)$$

The work in [11] approximates R_{TEG} based on the switching frequency f_s of the inductor-based boost converter as given in Eq. (2.6), where L is the inductor value. Hence, R_{in} is matched with R_{TEG} by varying f_s of the boost converter using a pulse frequency modulation (PFM) control.

$$R_{TEG} = \frac{2 \times L \times f_s}{D^2} \qquad (2.6)$$

Alternatively, a 3-dimensional (3D) MPPT is discussed in [14] to achieve impedance matching in a charge pump design. This is done by tuning three parameters of the charge pump: switching frequency f_s, capacitance C, and number of charge pump stages N.

2.2.4 Voltage Regulator

There are three main types of voltage regulators: LDO regulator, switched inductor (SI), and switched capacitor (SC). Each one is discussed in details in the coming sections.

2.2.4.1 LDO Regulator

Figure 2.7 shows the regulator circuit. It consists of a pass transistor M_p and negative feedback. The control feedback is implemented via the error amplifier and the voltage division. The output voltage level depends on the voltage drop across the transistor as well as the voltage division across R_1 and R_2 as given in Eq. (2.7). The output voltage will be compared to the reference voltage by the error amplifier

Fig. 2.7 LDO regulator circuit

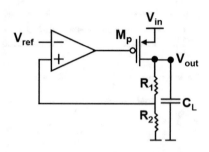

which in turn adjusts the gate of the pass transistor to maintain the desired output voltage.

$$V_{out} = V_{ref}(1 + \frac{R_1}{R_2})$$ (2.7)

The dropout voltage V_{do} across the transistor is defined as the minimum difference between the input voltage and the output voltage as given in Eq. (2.8).

$$V_{do} = V_{in} - V_{out}$$ (2.8)

The dropout voltage affects the power efficiency as given in Eq. (2.9), where I_q is the quiescent current consumed by the error amplifier, voltage divider, and voltage reference circuit. By observing Eq. (2.9), LDO regulator can achieve high power efficiency when both V_{do} and I_q are minimized. This means that the LDO regulator suffers from low power efficiency at high voltage conversion ratio [15]. However, it can generate a regulated output voltage with low ripple and it can be completely integrated on-chip with a small area.

$$\eta = \frac{P_{out}}{P_{in}} = \frac{V_{out} I_{out}}{V_{in} I_{in}} = \frac{(V_{in} - V_{do}) I_{out}}{V_{in}(I_{out} + I_q)}$$ (2.9)

2.2.4.2   Switched Inductor Voltage Regulator

Figure 2.8a shows the SI voltage regulator. It consists of two switches S_1 and S_2, inductor L, and load capacitor C_L. SI regulator has two different operation states which are on-state T_{on} and off-state T_{off}. During T_{on} where S_1 is on and S_2 is off as shown in Fig. 2.8b, both the inductor and the capacitor are charged by the input voltage so that the current through the inductor I_{in} flows into the output load. In addition, the output voltage is controlled by the duty cycle of the switch. Once V_{out} reaches the desired value, the SI regulator turns into off-state T_{off} where S_1 is off and S_2 is on. During T_{off} as shown in Fig. 2.8c, both the inductor and the load capacitor are discharged into the load such that I_{in} and V_{out} are decreased.

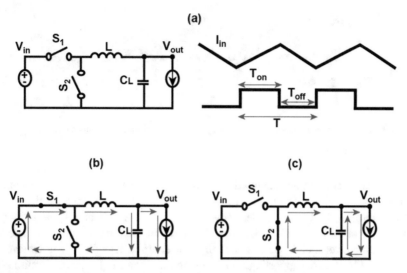

Fig. 2.8 (a) SI regulator circuit; (b) configuration of SI regulator during T_{on}; (c) configuration of SI regulator during T_{off}

The advantage of the SI voltage regulator is that it can provide high power efficiency up to 90% [16]. Low frequency (100 KHz to 3 MHz) SI regulator provides high power efficiency [15] but with an off-chip inductor (μH range) [17–19]. However, there are still many high frequency (100–600 MHz) SI regulators using very small on-chip (1–2 nH) and on-package inductors [20–22]. The main losses contributors in SI regulator are conduction and switching losses P_{cond} and P_{sw}, respectively [16]. P_{cond}, as given in Eq. (2.10), is proportional to the transistor drain–source resistance R_{ds} and to the quadratic drain–source current I_{ds}. Hence, this type of losses is dominant at heavy load current. In contrast, P_{sw} which is given in Eq. (2.11) is dominant at light load current where C_{ox} is the oxide capacitance since it depends on the switching frequency rather than the load current.

$$P_{cond} = I_{ds}^2 R_{ds} \qquad (2.10)$$

$$P_{sw} = C_{ox} V_{in}^2 f_s \qquad (2.11)$$

2.2.4.3 Switched Capacitor Voltage Regulator

Figure 2.9a shows the SC circuit. It consists of two switches, flying capacitor C_{fly}, and load capacitor C_L. The output voltage is generated by charging C_{fly} and discharging its energy into C_L as shown in Fig. 2.9b, c, respectively.

The output voltage level depends on the number of flying capacitors and how they are configured. This configuration allows providing a voltage gain g which is defined as the ratio of the output voltage at no load condition V_{NL} to the input voltage. The ideal power efficiency is given in Eq. (2.12) [23] where only the losses

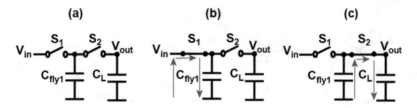

Fig. 2.9 (a) SC regulator circuit; (b) charging phase of SC regulator; (c) discharging phase of SC regulator

due to charge sharing are captured. This means that the closer the output voltage to the no load voltage, the higher the power efficiency. For example, if the gain is set to 1 and $V_{in} = 1.2$ V, the ideal power efficiency is 83.3% at $V_{out} = 1$ V while it is 58.3% at $V_{out} = 0.7$ V. Therefore, the SC circuit requires to switch to another suitable gain level such as 2/3 so that higher power efficiency of 87.5% is achieved at $V_{out} = 0.7$ V. Hence, supporting multiple gain topologies in the same circuit needs to be used in order to maximize the power efficiency at different output voltage levels.

$$\eta = \frac{V_{out}}{V_{NL}} = \frac{V_{out}}{g \times V_{in}} \tag{2.12}$$

As well as the SI voltage regulator, different power losses are associated with the SC voltage regulator which are: conduction losses, switching losses, flying capacitor losses, and bottom plate parasitic capacitance losses [24, 25]. Conduction and switching losses have been already discussed in the previous section and regulator. Flying capacitor losses P_{cfly} are defined as charge losses caused by C_{cfly} during the charge transfer through charging and discharging phases, as given in Eq. (2.13), where I_L is the load current.

$$P_{cfly} = \frac{I_L^2}{C_{fly} f_s} \tag{2.13}$$

Equation (2.13) shows that P_{cfly} is inversely proportional to the C_{fly} and f_s and proportional to quadratic I_L. This means that P_{cfly} is a load current dependent losses. On the other hand, P_{bp} as given in Eq. (2.14) is caused by the bottom plate capacitor C_{bp} such that $C_{bp} = \alpha C_{fly}$ where α depends on the technology used. C_{bp} is formed due to the PN junction between the N well and the P substrate.

$$P_{pb} = C_{bp} V_{in}^2 f_s \tag{2.14}$$

Table 2.1 summarizes the SC design parameters and their impact on the power losses. Switching frequency has an impact on all types of losses but in a different way. For example, as the switching frequency increases, the switching activities of the CMOS transistors increase which results in a faster charge transfer from the

Table 2.1 Comparison table between the SC design parameter and its impact on the power losses

Parameter	P$_{clfy}$	P$_{cond}$	P$_{sw}$	P$_{bp}$
f$_s$	Inversely Proportional	–	Proportional	Proportional
W	–	Inversely Proportional	Proportional	–
C$_{fly}$	Inversely Proportional	–	–	Proportional

Table 2.2 Comparison of different voltage regulator topologies

Comparison	LDO	SI	SC
Integration	On-chip	Off-chip	On-chip
Efficiency	Low*	High	Moderate
Speed	High	Low	Low

*At high voltage conversion

input to the output. This reduces the output voltage ripples and the flying capacitor losses. However, the switching losses and bottom plate parasitic losses will be increased due to the losses by gate drivers and parasitic capacitance, respectively [25]. On the other hand, the flying capacitor is proportional to the bottom plate parasitic losses since the bottom plate capacitor is affected by the total value of the flying capacitors. Yet, larger flying capacitor can store more charges with a high level of energy and discharge them into the load with low voltage ripples. This will reduce the flying capacitor losses. Moreover, the transistor width has opposite impact on both switching and conduction losses. All the parameters have to be optimized to achieve optimum power efficiency.

Comparing the SC regulator with the previous voltage regulators, it compromises the advantages of LDO and SI regulators where it can provide high power efficiency at different voltage conversion ratios (voltage gains) and can be implemented on-chip. However, as mentioned before, the SC regulator has inherent losses during the charge transfer. In addition, SC and SI suffer from high voltage ripple at low frequency. To minimize the voltage ripple, a large load capacitor can be added but it increases the active area of the regulator. An interleaving phase technique is another method to filter the voltage ripple in the switched regulators [24, 26]. Table 2.2 summarizes the advantages and disadvantages of LDO, SI, and SC voltage regulators.

2.3 Existed TEG-Based Power Management Solutions

Prior works have discussed different TEG-based PMU designs for low power devices. These designs are classified into two main categories: single stage PMU and cascaded two stages PMU. The TEG-based PMU design in [4] has a single stage inductor-based boost converter with TEG voltage level between 20 and 250 mV and generates a single load voltage level of 1 V. This PMU provides power efficiencies

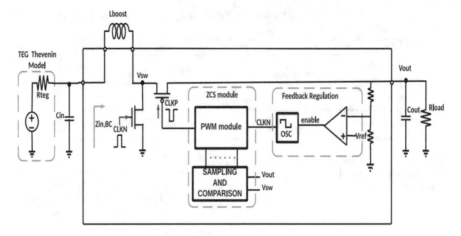

Fig. 2.10 Single stage TEG-based PMU using inductor-based boost converter [27]

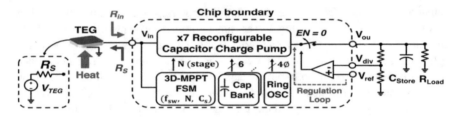

Fig. 2.11 Single stage TEG-based PMU using reconfigurable charge pump [14]

of 47% and 75% at load power levels of 24 μW and 175 μW, respectively. However, it does not support DVS. Similarly, the work in [27] as shown in Fig. 2.10 presents a single stage TEG-based PMU that consists of an inductor-based boost converter with a regulation feedback loop. The maximum efficiency obtained from postlayout simulation is 68% at a load power of 19 μW and TEG voltage of 22 mV.

The work in [14] implements another single stage PMU using a reconfigurable capacitor charge pump to boost the TEG input voltage from 0.27 V to generate a regulated voltage of 1 V as shown in Fig. 2.11. The PMU establishes a feedback regulation loop in order to support a regulated voltage level. The PMU delivers a load power up to 500 μW and provides a maximum power efficiency of 64% at a load power of 400 μA. The advantage of this PMU design is that the capacitors of the charge pump are fully integrated on-chip unlike the inductor in the inductor-based boost converter. However, this PMU design operates only when the TEG voltage is equal or greater than 0.27 V which is not suitable in wearable devices where TEG voltage level is less than 100 mV. On the other hand, Fig. 2.12 shows the block diagram of the cascaded two stages PMU that is powered by TEG and RF energy harvesting in [28]. The PMU design utilizes an inductor-based boost converter to boost up the TEG voltage from 30 mV to an unregulated voltage of

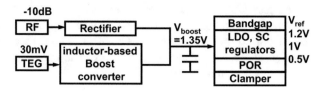

Fig. 2.12 Two stages PMU using inductor-based boost converter followed by SC and LDO regulators [28]

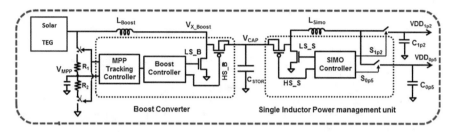

Fig. 2.13 Two stages PMU using inductor-based boost converter followed by a single inductor multiple output converter [29]

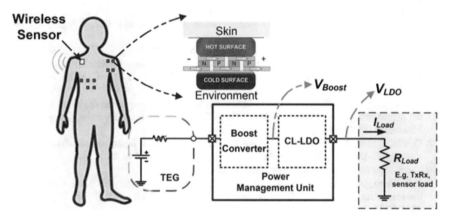

Fig. 2.14 Two stages TEG-based PMU using inductor-based boost converter followed by capacitor-less LDO (CL-LDO) regulator [30]

1.35 V. Voltage regulators of LDO and SC are used to generate different regulated voltage levels between 1.2 and 0.5 V.

The PMU design in [29] utilizes inductor-based boost converter for solar and TEG energy harvesting followed by single input multiple outputs (SIMO) to generate regulated voltage levels of 1.2 V and 0.5 V as shown in Fig. 2.13. The advantage of this design is that it provides high end-to-end efficiency of 74.9% (due to the use of SIMO) but at the cost of adding extra off-chip inductor. The TEG-based PMU in [30] consists of an inductor-based boost converter with a cascaded capacitor-less LDO (CL-LDO) regulator using a set of TEG arrays as shown in Fig. 2.14. The inductor-based converter is used as a front-end to boost the TEG voltage from 200 mV to 1.8 V. The CL-LDO is used as a back-end to generate a regulated and clean voltage level of 1.6 V. The maximum end-to-end power efficiency of this design is 57.57%.

References

1. D. Paul, "Thermoelectric energy harvesting," in *ICT-energy-concepts towards zero-power information and communication technology*. Intech, 2014.
2. D. Champier, "Thermoelectric generators: A review of applications," *Energy Conversion and Management*, vol. 140, pp. 167–181, 2017.
3. Y. Zhang, F. Zhang, Y. Shakhsheer, J. D. Silver, A. Klinefelter, M. Nagaraju, J. Boley, J. Pandey, A. Shrivastava, E. J. Carlson *et al.*, "A batteryless 19 µw MICS/ISM-band energy harvesting body sensor node SoC for ExG applications," *IEEE Journal of Solid-State Circuits*, vol. 48, no. 1, pp. 199–213, 2013.
4. E. J. Carlson, K. Strunz, and B. P. Otis, "A 20 mv input boost converter with efficient digital control for thermoelectric energy harvesting," *IEEE Journal of Solid-State Circuits*, vol. 45, no. 4, pp. 741–750, 2010.
5. Y. K. Ramadass and A. P. Chandrakasan, "A battery-less thermoelectric energy harvesting interface circuit with 35 mv startup voltage," *IEEE Journal of Solid-State Circuits*, vol. 46, no. 1, pp. 333–341, 2011.
6. J.-P. Im, S.-W. Wang, K.-H. Lee, Y.-J. Woo, Y.-S. Yuk, T.-H. Kong, S.-W. Hong, S.-T. Ryu, and G.-H. Cho, "A 40 mV transformer-reuse self-startup boost converter with MPPT control for thermoelectric energy harvesting," in *2012 IEEE International Solid-State Circuits Conference*. IEEE, 2012, pp. 104–106.
7. G. Pillonnet and T. Martinez, "Sub-threshold startup charge pump using depletion MOSFET for a low-voltage harvesting application," in *2015 IEEE Energy Conversion Congress and Exposition (ECCE)*. IEEE, 2015, pp. 3143–3147.
8. Q. Liu, X. Wu, M. Zhao, L. Wang, and X. Shen, "30–300 mV input, ultra-low power, self-startup DC–DC boost converter for energy harvesting system," in *2012 IEEE Asia Pacific Conference on Circuits and Systems*. IEEE, 2012, pp. 432–435.
9. P.-S. Weng, H.-Y. Tang, P.-C. Ku, and L.-H. Lu, "50 mv-input batteryless boost converter for thermal energy harvesting," *IEEE Journal of Solid-State Circuits*, vol. 48, no. 4, pp. 1031–1041, 2013.
10. C. Tse, S. Wong, and M. Chow, "On lossless switched-capacitor power converters," *Power Electronics, IEEE Transactions on*, vol. 10, no. 3, pp. 286–291, 1995.
11. S. Carreon-Bautista, A. Eladawy, A. N. Mohieldin, and E. Sanchez-Sinencio, "Boost converter with dynamic input impedance matching for energy harvesting with multi-array thermoelectric generators," *IEEE Transactions on Industrial Electronics*, vol. 61, no. 10, pp. 5345–5353, 2014.
12. M. Alhawari, B. Mohammad, H. Saleh, and M. Ismail, "An efficient zero current switching control for l-based DC–DC converters in TEG applications," *IEEE Transactions on Circuits and Systems II: Express Briefs*, vol. 64, no. 3, pp. 294–298, 2017.
13. M. Alhwari, "Multi-source energy harvesting interface circuits for biomedical wearable electronics," Ph.D. dissertation, Electrical and Computer Engineering, 2016.
14. S. Yoon, S. Carreon-Bautista, and E. Sánchez-Sinencio, "An area efficient thermal energy harvester with reconfigurable capacitor charge pump for IoT applications," *IEEE Transactions on Circuits and Systems II: Express Briefs*, 2018.

15. C.-W. Chen and A. Fayed, "A low-power dual-frequency SIMO buck converter topology with fully-integrated outputs and fast dynamic operation in 45 nm CMOS," *Solid-State Circuits, IEEE Journal of*, vol. 50, no. 9, pp. 2161–2173, 2015.
16. P.-J. Liu, W.-S. Ye, J.-N. Tai, H.-S. Chen, J.-H. Chen, and Y.-J. E. Chen, "A high-efficiency CMOS DC-DC converter with 9 μ s transient recovery time," *IEEE Transactions on Circuits and Systems I: Regular Papers*, vol. 59, no. 3, pp. 575–583, 2012.
17. D. Lu, Y. Qian, and Z. Hong, "4.3 An 87%-peak-efficiency DVS-capable single-inductor 4-output DC-DC buck converter with ripple-based adaptive off-time control," in *Solid-State Circuits Conference Digest of Technical Papers (ISSCC), 2014 IEEE International*. IEEE, 2014, pp. 82–83.
18. C. Tao and A. A. Fayed, "A low-noise PFM-controlled buck converter for low-power applications," *Circuits and Systems I: Regular Papers, IEEE Transactions on*, vol. 59, no. 12, pp. 3071–3080, 2012.
19. M. Belloni, E. Bonizzoni, E. Kiseliovas, P. Malcovati, F. Maloberti, T. Peltola, and T. Teppo, "A 4-output single-inductor DC-DC buck converter with self-boosted switch drivers and 1.2 a total output current," in *Solid-State Circuits Conference, 2008. ISSCC 2008. Digest of Technical Papers. IEEE International*. IEEE, 2008, pp. 444–626.
20. W. Kim, D. M. Brooks, and G.-Y. Wei, "A fully-integrated 3-level DC/DC converter for nanosecond-scale DVS with fast shunt regulation," in *Solid-State Circuits Conference Digest of Technical Papers (ISSCC), 2011 IEEE International*. IEEE, 2011, pp. 268–270.
21. J. Wibben and R. Harjani, "A high-efficiency DC–DC converter using 2 nH integrated inductors," *Solid-State Circuits, IEEE Journal of*, vol. 43, no. 4, pp. 844–854, 2008.
22. W. Fu and A. Fayed, "A self-regulated 588 MHZ buck regulator with on-chip passives and circuit stuffing in 65 nm," in *Circuits and Systems (MWSCAS), 2014 IEEE 57th International Midwest Symposium on*. IEEE, 2014, pp. 338–341.
23. Y. K. Ramadass, "Energy processing circuits for low-power applications," Ph.D. dissertation, MIT, 2009.
24. H.-P. Le, S. R. Sanders, and E. Alon, "Design techniques for fully integrated switched-capacitor DC-DC converters," *IEEE Journal of Solid-State Circuits*, vol. 46, no. 9, pp. 2120–2131, 2011.
25. M. D. Seeman, S. R. Sanders, and J. M. Rabaey, "An ultra-low-power power management IC for energy-scavenged wireless sensor nodes," in *2008 IEEE Power Electronics Specialists Conference*. IEEE, 2008, pp. 925–931.
26. N. Butzen and M. Steyaert, "12.2 a 94.6%-efficiency fully integrated switched-capacitor DC-DC converter in baseline 40 nm CMOS using scalable parasitic charge redistribution," in *Solid-State Circuits Conference (ISSCC), 2016 IEEE International*. IEEE, 2016, pp. 220–221.
27. V. Priya, M. K. Rajendran, S. Kansal, and A. Dutta, "A 11 mV single stage thermal energy harvesting regulator with effective control scheme for extended peak load," in *SoC Design Conference (ISOCC), 2016 International*. IEEE, 2016, pp. 113–114.
28. F. Zhang, Y. Zhang, J. Silver, Y. Shakhsheer, M. Nagaraju, A. Klinefelter, J. Pandey, J. Boley, E. Carlson, A. Shrivastava *et al.*, "A Batteryless 19 μW MICS/ISM-band Energy Harvesting Body Area Sensor Node SoC," in *Solid-State Circuits Conference Digest of Technical Papers (ISSCC), 2012 IEEE International*. IEEE, 2012, pp. 298–300.
29. A. Klinefelter, N. E. Roberts, Y. Shakhsheer, P. Gonzalez, A. Shrivastava, A. Roy, K. Craig, M. Faisal, J. Boley, S. Oh *et al.*, "21.3 a 6.45 μw self-powered IoT SoC with integrated energy-harvesting power management and ULP asymmetric radios," in *Solid-State Circuits Conference-(ISSCC), 2015 IEEE International*. IEEE, 2015, pp. 1–3.
30. J. Zarate-Roldan, S. Carreon-Bautista, A. Costilla-Reyes, and E. Sánchez-Sinencio, "A power management unit with 40 dB switching-noise-suppression for a thermal harvesting array." *IEEE Trans. on Circuits and Systems*, vol. 62, no. 8, pp. 1918–1928, 2015.

Chapter 3
TEG-Based Power Management Designs and Characterizations

This chapter presents a detailed characterization of several PMUs that target thermal energy harvesting wearable device. Several TEG-based PMU design options are proposed to select the appropriate option according to the blocks' requirements within the device given in Fig. 1.12. In this chapter, three various TEG-based PMUs are characterized: design 1: switched inductor (SI) boost converter followed by a switched capacitor (SC) voltage regulator, design 2: SI followed by LDO regulator, and design 3: SI followed by two voltage regulators in series of SC and LDO. The characterization is based on silicon measurements where the TEG-based PMUs are fabricated in a single chip in 65 nm CMOS. Power efficiency, area, and voltage ripple have been measured for different PMU topologies.

3.1 Proposed TEG-Based PMU Architecture

According to our target application as given in Fig. 1.12, different load voltages are required such as 1 V, 0.8 V, and 0.6 V in order to support different power domains. In our characterizations, the TEG-based PMU is designed to generate two load voltage levels V_L of 0.8 V and 0.6 V from TEG voltages V_{TEG} of 50 mV and 65 mV over a load current I_L of 10 μA to 100 μA.

Figure 3.1 shows three different TEG-based PMU design options using cascaded stages of DC–DC converters. The cascaded stages of PMU are chosen rather than a single stage in order to support DVS which is required in our system. The inductor-based boost converter (we call it in this section switched inductor (SI) boost converter) is employed since it provides high boost ratio and achieves high power efficiency, however, it needs off-chip inductor. Conversely, SC and LDO voltage regulators are chosen as second and third stage rather than SI step down converter because the former are completely integrated on-chip and it is preferable to minimize the number of the off-chip components in the PMU. Note that only the

© Springer Nature Switzerland AG 2020
D. Kilani et al., *Power Management for Wearable Electronic Devices*, Analog Circuits and Signal Processing, https://doi.org/10.1007/978-3-030-37884-4_3

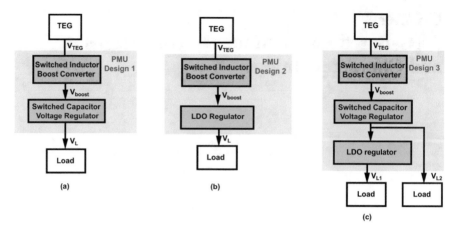

Fig. 3.1 Block diagram of TEG-based PMU designs using; (**a**) cascaded two stages of SI boost converter followed by SC voltage regulator; (**b**) cascaded two stages of SI boost converter followed by LDO regulator; and (**c**) cascaded three stages of SI boost converter followed by SC which is followed by LDO

SC circuit design represents the author's work in addition to the characterizations of these three TEG-based PMU designs. To startup the system given in Fig. 1.12, the PEG will be used to power up the digital logic of the boost converter. However, for the purpose of testing the system, external supplies are used to power the digital logic in the SI boost converter which are assumed to be available from the voltage regulators.

Figure 3.1a shows the first TEG-based PMU design option that consists of SI boost converter followed by SC regulator. The advantage of this design is that the SC converter can be configured to generate different conversion ratios to achieve high power efficiency at different input and load voltage levels as given in Eq. (2.12). However, this design option suffers from large voltage ripple which is not suitable to supply RF and analog blocks. Figure 3.1b shows the second TEG-based PMU design option that consists of SI boost converter followed by LDO regulator. Since the LDO provides low voltage ripple, this design is considered as a reduced ripple PMU recommended for RF applications. However, it suffers from low power efficiency at high voltage conversion ratio (V_{boost}/V_L). Figure 3.1c shows the third TEG-based PMU design option that consists of SI boost converter followed by two cascaded voltage regulators of SC and LDO. The advantage of this TEG-based PMU is that it can support two load voltages: one from LDO for RF and analog block and one from SC for digital blocks. Moreover, it combines the advantages of both design 1 (SI+SC) and design 2 (SI+LDO) by providing low voltage ripple and high power efficiency at high voltage conversion ratio. For example, the ideal power efficiency when directly regulating from 1.2 to 0.6 V using LDO regulator is 50%. However, having two voltage regulation steps when SC regulates from 1.2 to 0.8 V, followed by LDO to regulate from 0.8 to 0.6 V would maximize the power efficiency. Based on the author's knowledge, design 3 has not been characterized in the literature in 65 nm CMOS.

Fig. 3.2 A switched inductor boost converter with ZCS implementation [1, 2]

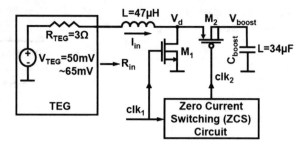

3.1.1 Switched Inductor Boost Converter Circuit Design

Figure 3.2 shows the SI boost converter. As depicted in the figure, M_1 is switched on during the charging phase so that the input current I_{in} that passes through the inductor is increased. Then, during the discharging phase, M_1 is switched off and M_2 is switched on so that the input current I_{in} is discharged into the output capacitor C_{boost} of the boost converter in order to generate the desired voltage V_{boost} that supplies the voltage regulator.

In our application where the system operates at light load power, the inductor current can go to negative values for certain periods which causes power loss and degrades the power efficiency. To minimize this loss, the SI boost converter can operate in discontinuous conduction mode (DCM) where the inductor current reaches zero. To detect the zero current crossing point, a zero current switching (ZCS) technique is implemented to turn off M_2 by clk_2 once the inductor current reaches zero as shown in Fig. 3.2, [1, 2]. This ZCS technique uses a coarse and fine delay steps to cover a wide delay range and a high resolution to locate the zero point of the inductor current using few control bits.

MPPT is employed by setting the switching frequency of the boost converter based on the value of R_{TEG} according to Eq. (2.6). The value of $R_{TEG} = 3\,\Omega$ is chosen based on the selected TEG devices G2-56-0570 from Tellurex that has 50 mV/C Seebeck effect [3]. Hence, the clock frequency is 65 kHz. The V_{TEG} range in the proposed design is between 50 and 65 mV to generate V_{boost} between 0.6 and 1.4 V. The inductor value is chosen to be 47 μH and C_{boost} is chosen to be 34 μF. The SI boost converter has been designed, implemented, and adopted from this thesis [4].

3.1.2 Switched Capacitor Regulator Circuit Design

Figure 3.3 shows the block diagram of the SC voltage regulator. It consists of a reconfigurable SC circuit, digital gain controller, pulse frequency modulation (PFM), and non-overlapping two- phase circuit. The reconfigurable SC circuit as shown in Fig. 3.4a consists of 3 flying capacitors C_{fly1-3} and 13 switches S_{1-13}.

Fig. 3.3 SC regulator block diagram [5]

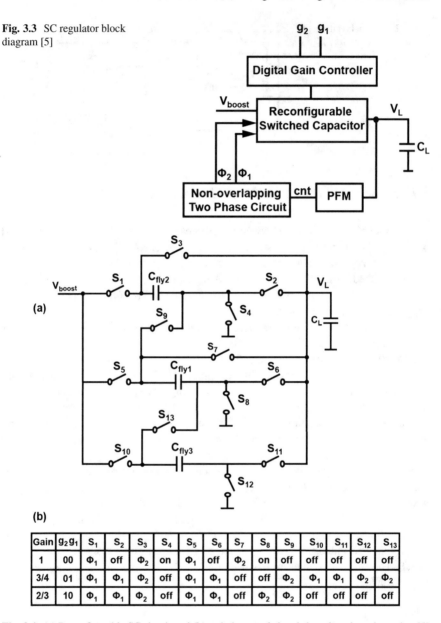

(a)

(b)

Gain	$g_2 g_1$	S_1	S_2	S_3	S_4	S_5	S_6	S_7	S_8	S_9	S_{10}	S_{11}	S_{12}	S_{13}
1	00	Φ_1	off	Φ_2	on	Φ_1	off	Φ_2	on	off	off	off	off	off
3/4	01	Φ_1	Φ_1	Φ_2	off	Φ_1	Φ_1	off	off	Φ_2	Φ_1	Φ_1	Φ_2	Φ_2
2/3	10	Φ_1	Φ_1	Φ_2	off	Φ_1	Φ_1	off	Φ_2	Φ_2	off	off	off	off

Fig. 3.4 (**a**) Reconfigurable SC circuit and (**b**) switch control signals based on the gain setting [5]

The SC regulator circuit provides multiple voltage gain settings g of 1, 3/4, and 2/3 controlled by two digital bits of g_1 and g_2 using digital gain controller. Each gain setting has a certain switch configuration controlled by the two clocks ϕ_1 and ϕ_2 as shown in Fig. 3.4b.

In addition, a PFM controller is utilized to regulate the load voltage V_L against the variation of the load current I_L. Such controller adjusts the switching frequency cnt based on the load current. If the load current increases, this means that the load voltage drops and therefore, the PFM increases the switching frequency to achieve the desired load voltage. On the other hand, if the load current decreases, the load voltage increases. As a result, the PFM reduces the switching frequency to reduce the load voltage according to the desired value. The value of each flying capacitor is chosen to be 225 pF and the value of load capacitor is 1.6 nF. Both flying and load capacitor are fully integrated on-chip. The maximum clock frequency is 13 MHz and each switch has a transistor width of 100 μm. The SC converter has been designed and implemented previously by the first author in her master thesis in [6].

3.1.3   LDO Regulator Circuit Design

Figure 3.5 shows the circuit design of the LDO regulator. It is a less conventional circuit as the classical OPAMP topology is replaced by a differential pair and a voltage tracking loop. The transistors M1 and M2 work as the error amplifier forcing the load voltage V_L to equal the reference voltage V_{ref}. M4 is the series power MOSFET supplying current to the output. The voltage tracking loop operates as follows. The current I_0 is forced in transistor M2 through M3 and M4. In consequence, the current in M1 is $2I_0 - I_0 = I_0$. By neglecting the early effect due to unequal drain–source voltages, we can infer that $V_{gs(M1)} = V_{gs(M2)}$ implying $V_{ref} = V_L$. By adding a resistor between the drain of M1 and V_{boost}, the drain–

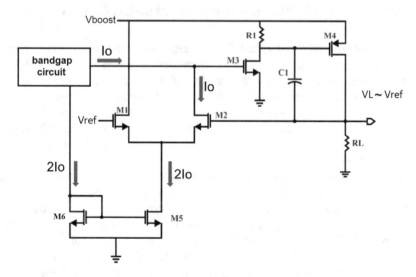

Fig. 3.5 LDO regulator circuit [7]

Table 3.1 Transistors' size
of the LDO regulator

Transistor	Size
M_1, M_2	150 nm/100 nm
M_3	150 nm/10 μm
M_4	100 μm/200 nm
M_5, M_6	1 μm/300 nm

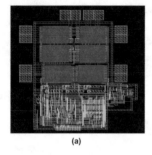
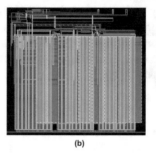
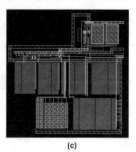

(a)   (b)   (c)

Fig. 3.6 Layout of (**a**) SI boost converter; (**b**) SC buck converter; (**c**) LDO

source voltages of M1 and M2 could be made equal. In consequence, the currents of
M1 and M2 will become equal. This, then, will improve the tracking of V_{ref} at the
output V_L. Capacitor C1 bypasses transistor M4 at high frequencies and stabilizes
the loop. Table 3.1 shows the transistors' size of the LDO regulator. The LDO has
been designed, implemented, and adopted from this thesis [7].

3.2 TEG-Based PMU Measurements in 65 nm CMOS

The SI boost converter, LDO, and SC regulators have been fabricated in 65 nm
CMOS. The layout of the three DC–DC converters is shown in Fig. 3.6. The die
photo is shown in Fig. 3.7. The active area of the SI converter, LDO, and SC
regulators are 0.0360 mm², 0.0357 mm², and 0.495 mm², respectively. Figure 3.8
shows the test setup used to characterize the performance of various TEG-based
PMU designs. The setup includes a DC power analyzer to emulate the TEG source
and to measure the end-to-end efficiency. Further, an oscilloscope is used to measure
the time domain signals.

Figure 3.9 shows the measured power efficiency of the stand-alone SI boost
converter, SC, and LDO regulators. Figure 3.9a shows the measured power efficiency of the SI boost converter at different load currents and $V_{TEG} = 50$ mV and
65 mV. The power efficiency is measured by dividing the output power over the
input power. Note that the input power is fixed for a certain TEG voltage value due
to the MPPT technique. The peak efficiency reaches 70% at $V_{TEG} = 65$ mV. The
SI boost efficiency changes dramatically according to the load current. Figure 3.9b
indicates the output voltage of the SI boost converter at $V_{TEG} = 50$ mV and 65 mV

Fig. 3.7 Die photograph of TEG-based PMU

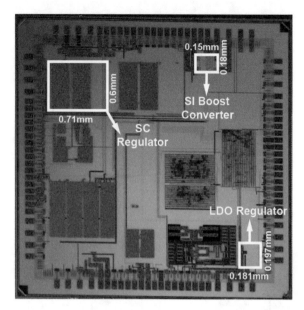

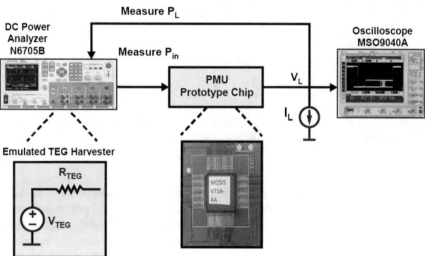

Fig. 3.8 Physical setup for characterizing different PMU designs

which is considered as the input voltage to the voltage regulator blocks. As depicted in the figure, the output voltage of the SI boost converter is not regulated as it varies according to the load current. Figure 3.9c shows the measured power efficiency of SC regulator supplied by a fixed input voltage of 1.2 V from the DC power analyzer at different load voltages of $V_L = 1$ V, 0.8 V, and 0.6 V. The peak efficiency of 81% is achieved at $V_L = 1$ V and $I_L = 500\,\mu$A. The SC regulator supports a load

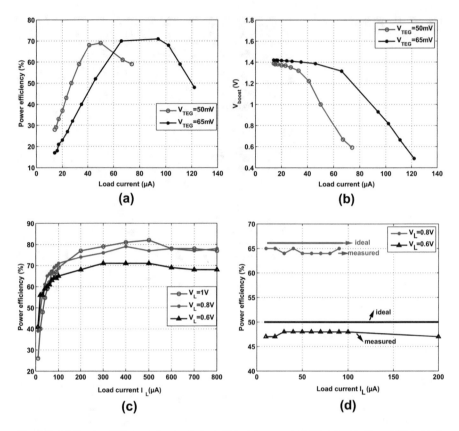

Fig. 3.9 (**a**) Power efficiency of a stand-alone SI boost converter; (**b**) output voltage of the stand-alone SI boost converter; (**c**) power efficiency of the stand-alone SC supplied by 1.2 V; (**d**) power efficiency of LDO supplied by 1.2 V

current up to $800\,\mu\mathrm{A}$. Figure 3.9d shows the measured power efficiency of the LDO regulator supplied by a fixed input voltage of 1.2 V from the DC power analyzer at different load voltages $V_L = 0.8$ V and 0.6 V. As depicted in the figure, the power efficiency is steady over the load current variations. The LDO regulator is capable of providing a load current up to $200\,\mu\mathrm{A}$ at $V_L = 0.6$ V. The LDO has better power efficiency than the SC regulator at low load current of $10\,\mu\mathrm{A}$ because it has lower measured quiescent current of $0.6\,\mu\mathrm{A}$ compared to $5\,\mu\mathrm{A}$ for SC regulator.

Figure 3.10 shows the power efficiency, V_{boost} and V_L of the cascaded TEG-based PMU design 1 (SI+SC), design 2 (SI+LDO), and design 3 (SI+SC+LDO) at $V_{TEG} = 50\,\mathrm{mV}$ and $V_L = 0.6$ V. The SC regulator utilizes a gain of 2/3 to regulate the output voltage at 0.6 V. As depicted in the figure, the various PMU designs are capable to regulate and support a load voltage of 0.6 V even when V_{boost} is not constant. At the load current range of 10–$30\,\mu\mathrm{A}$, SI+SC+LDO design has slightly higher power efficiency than the rest because the output voltage is higher. The output

Fig. 3.10 (**a**) End-to-end power efficiency at $V_{TEG} = 50\,\text{mV}$ and $V_L = 0.6\,\text{V}$; (**b**) input voltage versus load current; and (**c**) load voltage versus load current. The gain of the SC $= 2/3$

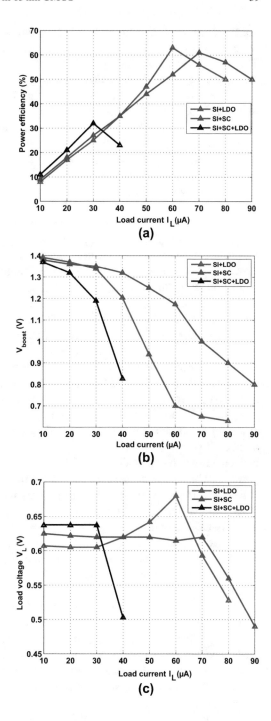

voltage of the PMU design that is driven by the LDO regulator has 50 mV difference from the reference voltage if the input voltage of the LDO is between 0.7 and 0.8 V. This is because at low supply voltage, the transistors of M4 and M5 in the LDO are no longer in the saturation region and operate in the linear region. This can be also observed at $I_L = 60 \, \mu A$ where V_L that is supported by SI+LDO is greater than 0.65 V because of the low V_{boost} of 0.7 V. This makes the peak efficiency 62% for SI+LDO design at $I_L = 60 \, \mu A$ which is higher than the power efficiency of SI+SC design by 12%. As the load current increases, and as long as the output power is less than the fixed input power, the power efficiency always increases. Once the output power exceeds the input power, the regulation cannot be maintained and the load voltage will stop the regulation at a certain load current point. As shown in Fig. 3.10c, the voltage in SI+SC design can be regulated at a wider load current range up to $70 \, \mu A$ because it is more efficient whereas the load current range decreases to $50 \, \mu A$ and $30 \, \mu A$ for SI+LDO and SI+SC+LDO, respectively. Figure 3.11 indicates the measured power efficiency, V_{boost} and V_L versus load current at $V_{TEG} = 50 \, mV$ and $V_L = 0.8 \, V$. The SI+LDO design has higher voltage regulation range, in this case, than SI+SC design. This is because the gain used for the SC to generate 0.8 V is 3/4 and cannot provide $V_L = 0.8 \, V$ if $V_{boost} < 1.1 \, V$. For example, at $I_L = 45 \, \mu A$ in SI+SC design, $V_{boost} = 1 \, V$ and with a gain of 3/4, V_L drops to 0.75 V which cannot support the desired $V_L = 0.8 \, V$. On the other hand, at $I_L = 45 \, \mu A$ in SI+LDO design, $V_{boost} = 1.1 \, V$ which can provide $V_L = 0.8 \, V$. For the same reason, SI+LDO design has more current load capability up to $60 \, \mu A$ than SI+SC and SI+SC+LDO design. Figures 3.12 and 3.13 show the measured power efficiency, V_{boost} and V_L versus load current when $V_{TEG} = 65 \, mV$ at $V_L = 0.6 \, V$ and $V_L = 0.8 \, V$, respectively. It can be obviously seen that the power efficiency is constant among the three TEG-based PMU designs regardless of the number of power stages are being used. For a given load current, the output power is fixed among the three TEG-based PMUs. Therefore, the efficiency at fixed input power will be always the same as long as the output is regulated. However, there are still some differences among the three designs such as the output voltage regulation range and load current capability as discussed previously. The gain of the SC in the SI+SC+LDO design is changed to 1 as shown in Fig. 3.13 which supports wider load current up to $100 \, \mu A$ compared to SI+SC+LDO when the gain of SC = 3/4. This is because the SC is capable to regulate the load voltage even when V_{boost} is decreased below 1 V unlike in the gain of 3/4 where SI+SC+LDO can support V_L when only V_{boost} is above or equal 1.2 V.

Figure 3.14 shows the measured time domain waveforms of the three PMU designs at $V_{TEG} = 50 \, mV$, $V_L = 0.6 \, V$, and $I_L = 10 \, \mu A$. As depicted in the figure, the three TEG-based PMU designs are capable to regulate a load voltage of 0.6 V. The TEG-based PMU that has the LDO regulator as the last stage has lower voltage ripple of 12 mV compared to 35 mV in the design where SC drives the load. Table 3.2 shows a comparison between our TEG-based PMUs and the state of the art designs in the literature. Our TEG-based PMUs support DVS that can provide multiple regulated load voltages of 0.8 V and 0.6 V unlike the single stage PMU in [8–10] where it supports single voltage level. Comparing our two stages PMU of

Fig. 3.11 (**a**) End-to-end power efficiency at $V_{TEG} = 50\,\text{mV}$ and $V_L = 0.8\,\text{V}$; (**b**) input voltage versus load current; and (**c**) load voltage versus load current. The gain of the $= 3/4$

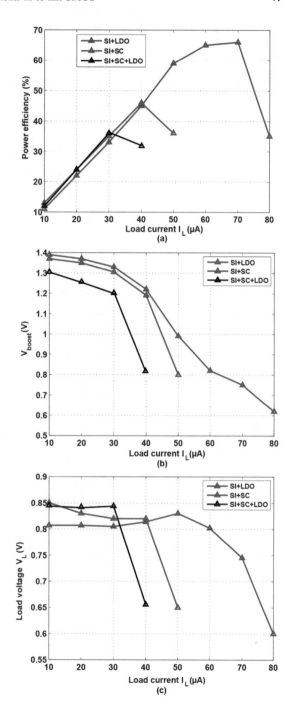

Fig. 3.12 (a) End-to-end power efficiency at $V_{TEG} = 65\,\text{mV}$ and $V_L = 0.6\,\text{V}$; (b) input voltage versus load current; and (c) load voltage versus load current. The gain of the SC = 2/3

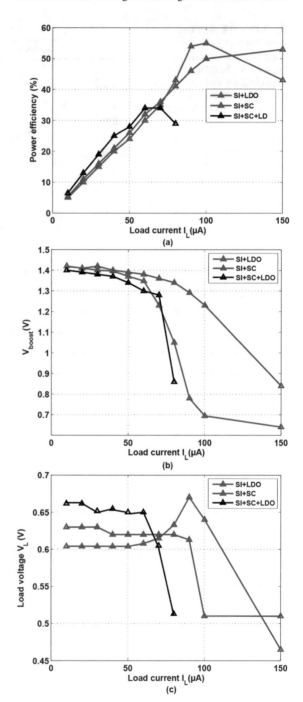

Fig. 3.13 (**a**) End-to-end
power efficiency at
$V_{TEG} = 65\,\text{mV}$ and
$V_L = 0.8\,\text{V}$; (**b**) input voltage
versus load current; and (**c**)
load voltage versus load
current. The gain of the SC in
SI+SC+LDO is 3/4 or 1. The
gain of the SC in the SI+SC
is 3/4

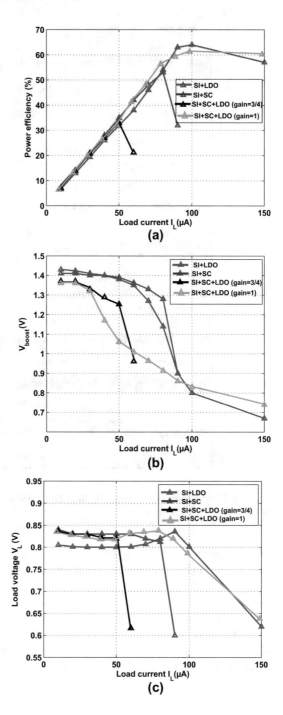

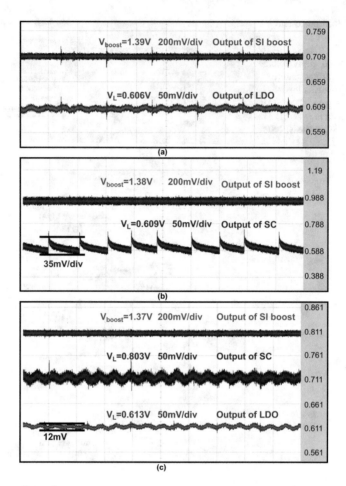

Fig. 3.14 Time domain waveforms of (**a**) SI+LDO; (**b**) SI+SC; and (**c**) SI+SC+LDO at $V_{TEG} = 50$ mV, $V_L = 0.6$ V and $I_L = 10$ μA

SI+LDO with the SI+LDO in [11], we achieve a higher power efficiency of 65% unlike 57% in [11]. Although the maximum power efficiency of SI+SIMO in [12] is 74.9% which is higher than 60% in our SI+SC design, it needs two off-chip inductors. Our SI boost converter has an area of 0.036 mm^2 and requires off-chip inductor. The maximum power efficiency of SI+SC+LDO is 60% where the gain of the SC used in this case is 1 to support higher load current at $V_L = 0.8$ V. The SC regulator occupies the largest area of 0.495 mm^2 whereas the LDO regulator occupies the smallest area of 0.0357 mm^2.

Table 3.2 Comparison table between our work of TEG-based PMU and prior works

	[8]	[9]	[10]	[11]	[12]	This Work		
Technology	130 nm CMOS	180 mn UMC	130 nm CMOS	500 nm CMOS	130 nm CMOS	65 nm CMOS	65 nm CMOS	65 nm CMOS
Topology	Single SI boost	Single SI boost	Single charge pump	Two stages SI boost+LDO	SI boost+SIMO	Two stages SI boost+SC	Two stages SI boost+LDO	Three stages SI boost+SC +LDO
VTEG	20 mV to 200 mV	11 mV to 100 mV	0.27 V to 1 V	50 mV-250 mV	10 mV	50 mV, 65 mV	50 mV, 65 mV	50 mV, 65 mV
VL	1 V	1 V	1 V	1.6 V	1.2 V, 0.5 V	0.8 V, 0.6 V	0.8 V, 0.6 V	0.8 V, 0.6 V
ηmax	47% @PL=24 µW VTEG=20 mV 75% @ PL=175 µW VTEG=100 mV	68% @PL=20 µW VTEG=22 mV	64% [a]	57%	74.9% @PL=100 µW VL=1.2 V 65.7% @ PL=100 µW VL=0.5 V	60% [b] @IL=70 µA, VL=0.6 V VTEG=50 mV	65% @IL=60 µA VL=0.8 V VTEG=50 mV	60% [c] @IL=90 µA VL=0.8 V VTEG=65 mV
Voltage ripple	20 mV	Not reported	80 mV	1 mV	Not reported	35 mV	12 mV	12 mV
Area	0.117 mm²	0.047 mm²	0.835 mm²	0.93 mm²	Not reported	0.531 m²	0.0717 mm²	0.566 mm²
DVS	No	No	No	No	Yes	Yes	Yes	Yes

[a]Postlayout simulation   [b]Gain of SC=2/3   [c]Gain of SC=1

Table 3.3 PMU design option usage based on the design requirement

Design requirement	PMU design option
Low voltage ripple	SI+LDO
Analog blocks	
High conversion ratio	SI+SC
Low output voltage level	
Digital blocks	
Multiple output voltages	SI+SC+LDO
Analog and digital blocks	

3.3 Summary and Recommendations

This chapter has presented silicon characterizations of three various TEG-based PMU designs targeting μW electronic devices. The TEG-based PMU consists of two parts: power conversion and power regulation. In the power conversion, SI boost converter is utilized to extract the maximum power point and boost the TEG voltage from 50 to 65 mV into 0.6 to 1.4 V. In contrast, the power regulation block regulates the boosted voltage into two load voltages of 0.6 V and 0.8 V in order to achieve voltage regulation and scaling over load currents of 10–100 μA. The three TEG-based PMU designs of SI+SC, SI+LDO, and SI+SC+LDO have similar power efficiencies at a certain load power because the input power from the SI is fixed due to MPPT. The peak efficiency of the three TEG-based PMUs fluctuates between 60 and 65% at different TEG and load voltages. The voltage regulation range of the SC is affected by both the input voltage and the voltage gain. Table 3.3 summarizes the selection of the PMU design option according to the block requirement within the device. SI+LDO PMU design is a good option for low voltage ripple which is suitable for analog and RF blocks. Conversely, SI+SC PMU design is preferable to support digital blocks since it can provide high voltage conversion ratio. Finally, SI+SC+LDO can support two output voltage levels from both SC and LDO regulators which can be used in both analog and digital blocks.

References

1. M. Alhawari, B. Mohammad, H. Saleh, and M. Ismail, "An efficient zero current switching control for L-based DC–DC converters in TEG applications," *IEEE Transactions on Circuits and Systems II: Express Briefs*, vol. 64, no. 3, pp. 294–298, 2017.
2. M. Alhawari, B. Mohammad, H. Saleh, and M. Ismail, "An Efficient Polarity Detection Technique for Thermoelectric Harvester in L-based Converters," *IEEE Transactions on Circuits and Systems I: Regular Papers*, vol. 64, no. 3, pp. 705–716, 2017.
3. Y. Ramadass. G2-56-0570 Thermoelectric Power Generation Module Specifications.
4. M. Alhwari, "Multi-source energy harvesting interface circuits for biomedical wearable electronics," Ph.D. dissertation, Electrical and Computer Engineering, 2016.
5. D. Kilani, M. Alhawari, B. Mohammad, H. Saleh, and M. Ismail, "An Efficient Switched-Capacitor DC-DC Buck Converter for Self-Powered Wearable Electronics," *IEEE Transactions on Circuits and Systems I: Regular Papers*, vol. 63, no. 10, pp. 1557–1566, 2016.
6. D. Kilani, "An efficient on-chip switched-capacitor DC-DC converter for ultra-low power applications," Master's thesis, Electrical and Computer Engineering, 2015.
7. H. Alblooshi, "Hybrid energy harvester for ultra low radios (ULP) applied to internet of things (IoT) and hybrid invisible tags," Master's thesis, Khalifa University of Science and Technology, 2016.
8. E. J. Carlson, K. Strunz, and B. P. Otis, "A 20 mV input boost converter with efficient digital control for thermoelectric energy harvesting," *IEEE Journal of Solid-State Circuits*, vol. 45, no. 4, pp. 741–750, 2010.
9. V. Priya, M. K. Rajendran, S. Kansal, and A. Dutta, "A 11 mV single stage thermal energy harvesting regulator with effective control scheme for extended peak load," in *SoC Design Conference (ISOCC), 2016 International*. IEEE, 2016, pp. 113–114.
10. S. Yoon, S. Carreon-Bautista, and E. Sánchez-Sinencio, "An area efficient thermal energy harvester with reconfigurable capacitor charge pump for IoT applications," *IEEE Transactions on Circuits and Systems II: Express Briefs*, 2018.
11. J. Zarate-Roldan, S. Carreon-Bautista, A. Costilla-Reyes, and E. Sánchez-Sinencio, "A power management unit with 40 dB switching-noise-suppression for a thermal harvesting array," *IEEE Trans. on Circuits and Systems*, vol. 62, no. 8, pp. 1918–1928, 2015.
12. A. Klinefelter, N. E. Roberts, Y. Shakhsheer, P. Gonzalez, A. Shrivastava, A. Roy, K. Craig, M. Faisal, J. Boley, S. Oh *et al.*, "21.3 a 6.45 μw self-powered IoT SoC with integrated energy-harvesting power management and ULP asymmetric radios," in *Solid-State Circuits Conference-(ISSCC), 2015 IEEE International*. IEEE, 2015, pp. 1–3.

Chapter 4
Dual-Outputs Switched Capacitor Voltage Regulator

Since multiple voltage levels are required simultaneously by the proposed wearable device in Fig. 1.12, multiple voltage regulators are needed. The switched capacitor (SC) voltage regulator is widely utilized in low power applications since the small capacitance can be fully integrated on-chip while driving μA current. Conventionally, multiple SC voltage regulators are required to generate simultaneous multiple regulated voltage levels. This way, each block in the chip operates at an optimum voltage level. However, this conventional PMU design suffers from a large area overhead because it is dominated by the area of the SC regulators [1]. The number of SC regulators increases as the number of the required regulated output voltage levels increases. Hence, more passive components (flying and load capacitors) are needed which makes the PMU area large. As a result, it is imperative to find an efficient way to generate multiple output voltages with minimum area overhead that targets μW power devices.

In this chapter, first, we discuss the state of the art multi-outputs SC regulators available in the literature. Then, we present our proposed dual-outputs SC regulator design followed by measurement results in 65 nm CMOS technology.

4.1 State of the Art Dual-Outputs Switched Capacitor Voltage Regulator

Several methods for generating simultaneous multiple output voltages have been introduced within the literature. The reported methods can be classified into three groups: parallel SC regulators [2], capacitor and switch sharing technique [1, 3–5], and time multiplexing technique [6, 7]. The work in [2] presents a triple-outputs PMU as illustrated in Fig. 4.1a. The design consists of three SC regulators: a binary reconfigurable SC and two parallel regulators of a Dickson charge pump with a voltage conversion ratio of 3/1 and a regulator with a voltage conversion ratio of

© Springer Nature Switzerland AG 2020
D. Kilani et al., *Power Management for Wearable Electronic Devices*, Analog Circuits and Signal Processing, https://doi.org/10.1007/978-3-030-37884-4_4

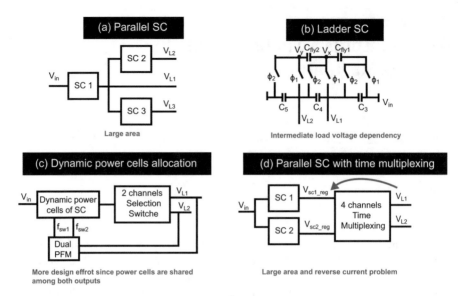

Fig. 4.1 (**a**) Parallel SC regulators, (**b**) ladder SC regulator, (**c**) dynamic power cell allocations; (**d**) parallel SC with time multiplexing

1/2. The three SC regulators utilize three independent frequency modulations to regulate the three output voltages over a wide range of load current. However, this conventional design suffers from a large area overhead since three regulators are used.

The second method is capacitor and switch sharing. Ladder SC shares the capacitor and the control switches to generate multiple outputs [3–5]. The work in [3] presents a ladder SC regulator that generates two output voltage levels simultaneously without changing the circuit topology as depicted in Fig. 4.1b. The output voltage levels equal to $2/3V_{dd}$ and $1/3\,V_{dd}$. The main drawback of this topology is that it provides outputs at fixed ratio which means that if the input voltage changes, then, the output voltage changes because it depends on the intermediate node voltage. Thus, ladder SC is not suitable for applications where stringent regulation is needed. This problem has been addressed in [4] where a ladder SC is followed by a differential inductor-based DC–DC buck converter. The triple-outputs ladder SC utilizes three inductor-based buck converters with a pulse width modulation (PWM) to regulate the intermediate node against the input voltage variation. Furthermore, to reduce the cross-regulation among the three output voltages due to load variation, a feedforward loop consisting of an ADC and a cross-regulation suppression block has been implemented. When the transient load occurs at one of the outputs, the cross-regulation suppression block adjusts the duty ratio of the switches that controls the other two outputs to minimize the cross-regulation. The drawback of this design is using three off-chip inductors. Another triple-outputs ladder SC design followed by the inductor-based buck converter is explained in [5]. The design utilizes an LC filter to reduce the output voltage ripple. In addition, a feedback loop is applied to regulate the output voltages. The work presented in [1] utilizes the flying capacitance and switch sharing method to improve the area. The design is implemented using 2 flying capacitors and 8 switches to achieve simultaneous voltage conversion ratios rather than using 3 capacitors and 11 switches in the individual designs. A sub-harmonic adaptive on-time generator is used to regulate two outputs and to adjust the on-time of the switches that control the charge transfer between the outputs to reduce the output voltage ripple. Five different clock signals are needed, rather than two clock signals in the conventional design.

The third method to generate multiple output voltage levels is the time multiplexing technique. The symmetrical dual-output SC regulator presented in [8] uses time multiplexing in addition to dynamic power cell allocations as depicted in Fig. 4.1c. However, this circuit requires more design effort since 82 power cells are shared between the two outputs. Thus, accurate timing and control circuits as wells as clock generators are needed to select and enable multiple SC power cells to support the two outputs based on the load current. This design generates two output voltages with the same value which might not be useful in devices where different voltage levels are required. The design in [6] introduces a dual-output SC regulator using time multiplexing and capacitance sharing. The flying capacitors are shared and the clock signals are time multiplexed to provide different conversion ratios of 1/3 and 2/3 which utilize two flying capacitors rather than four flying capacitors in the two parallel SCs. However, this design requires four clock signals rather than two clock signals in the two parallel SCs. In addition, 100 nF off-chip flying capacitors are used in this design.

The work in [7] as shown in Fig. 4.1d introduces two parallel SCs. The output voltages are generated through a 4-way channel switch that is controlled by the two non-overlapping phase clocks. The channel switch selection is based on the required load current which in turn is determined by the feedback loop. Both SCs regulate the two outputs for heavy load currents, whereas only SC1 supports the two outputs for light load currents. This solution suffers from reverse current issue when switching from a low to a high output voltage level. For example, if only SC1 is operating and V_{L2} is less than V_{L1}, then, if V_{L2} is achieved, whereas V_{L1} drops below the desired value, SC1 switches from V_{L2} into V_{L1}. Since $V_{L1} > V_{sc1-reg}$, the current flows backward, causing a voltage droop at V_{L1}. This droop is due to the charge sharing that occurs between the capacitors across the two nodes of $V_{sc-reg1}$ and V_{L1}. The reverse current problem is significant in applications where the power supplies are time multiplexed as discussed in [9]. The problem of the voltage droop is critical because if the load voltage drops below the minimum operating voltage, it can degrade the performance and might even results in a system failure. Therefore, the demand for a multi-outputs SC with no reverse current problem is increasing.

4.2   Proposed Dual-Outputs Switched Capacitor Voltage Regulator

Figure 4.2a illustrates the block diagram of the proposed PMU that generates two voltages of 1 V and 0.55 V. The PMU design consists of a single unit dual-outputs switched capacitor (DOSC) voltage regulator powered by an energy source to supply two units. The DOSC simultaneously provides two output voltage levels. In our system, the first load voltage $V_{L1} = 1$ V is used to support the heavy load block of unit 1 and the second load voltage $V_{L2} = 0.55$ V supports the light load block of unit 2. Figure 4.2b depicts the block diagram of the proposed DOSC regulator, where multiple voltage domains are achieved using the adaptive time multiplexing (ATM) technique.

To better explain the proposed design, Fig. 4.3 shows the block diagram of the DOSC regulator. The proposed DOSC consists of a SC regulator, a digital gain control, a deadtime phase generator, a pulse frequency modulation (PFM), and an ATM controller. The SC regulator has multiple gain settings, which are controlled by the digital gain bit g [10]. It provides a gain of 1 and 1/2 when $g = 1$ and $g = 0$, respectively, to achieve high power efficiency at different voltage levels. The deadtime phase generator prevents any overlap between ϕ_1 and ϕ_2 to reduce the short circuit current during the switching time between the two phases. The PFM regulates the output voltage $V_{sc\text{-}reg}$ over a wide range of load currents by modulating the switching frequency f_{sw}. Thus, the proposed DOSC regulator reduces the cross-regulation problem because the PFM feedback loop maintains, stabilizes, and

Fig. 4.2 (a) Block diagram of the proposed PMU using dual-outputs SC (DOSC) voltage regulator, (b) block diagram of the proposed DOSC including SC circuit, PFM, and adaptive time multiplexing

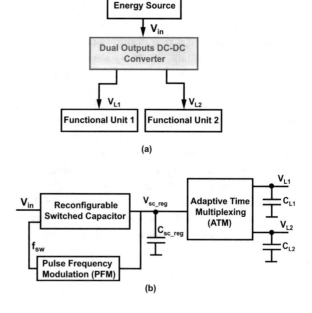

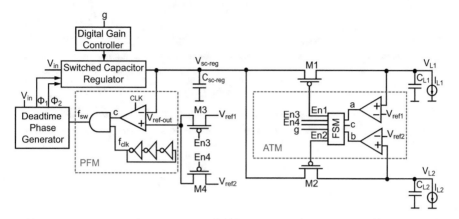

Fig. 4.3 Proposed dual-outputs switched capacitor voltage regulator

regulates the intermediate node voltage of $V_{sc\text{-}reg}$. The ATM controller generates and switches between the two load voltages and eliminates the reverse current problem. This scheme consists of two comparators and a finite state machine (FSM). The comparators compare V_{Ln} and $V_{refn}(n = 1, 2)$ and based on the comparison, the FSM determines the required switching voltage domain. The FSM controls $M1, M2, M3, M4$, and g based on not only the output comparison between V_{Ln} and V_{refn}, but also between $V_{sc\text{-}reg}$ and $V_{ref\text{-}out}$. This is important to eliminate the reverse current as will be discussed later. $M1$ and $M3$ control V_{L1}, whereas $M2$ and $M4$ control V_{L2}. Initially, the voltage level of $V_{sc\text{-}reg}$, V_{L1}, and V_{L2} is 0 V where the load capacitors $C_{sc\text{-}reg}$, C_{L1}, and C_{L2} are depleted. The highest priority of DOSC is given to generate the lower voltage $V_{L2} = 0.55$ V. Thus, the FSM generates a zero value for $En2$ and $En4$ to turn on $M2$ and $M4$, respectively. Subsequently, $En1$ and $En3$ are set to 1, which turns off $M1$ and $M3$, respectively, as shown in Fig. 4.4. In addition, the FSM sets the SC to a gain of 1/2. Note that in this case, $V_{sc\text{-}reg}$ and $V_{ref\text{-}out}$ are equal to $V_{ref2} = 0.55$ V. $M2$ and $M4$ remain on as long as C_{L2} is depleted. Once the desired voltage of 0.55 V is reached, the FSM switches off $M2$ and $M4$ ($En2 = En4 = 1$) and switches on $M1$ and $M3$ ($En1 = En3 = 0$) to enable a higher output voltage $V_{L1} = 1$ V as shown in Fig. 4.4. Moreover, the gain setting of SC is changed to 1. In this case, $V_{sc\text{-}reg}$ and $V_{ref\text{-}out}$ are equal to $V_{ref_1} = 1$ V.

4.2.1   Reconfigurable SC Circuit Design

The reconfigurable SC regulator consists of 2 flying capacitors and 8 switches as shown in Fig. 4.5a. The SC regulator is configured in such a way as to support two different voltage gain settings of 1 and 1/2. Figure 4.5b illustrates ϕ_1 and ϕ_2

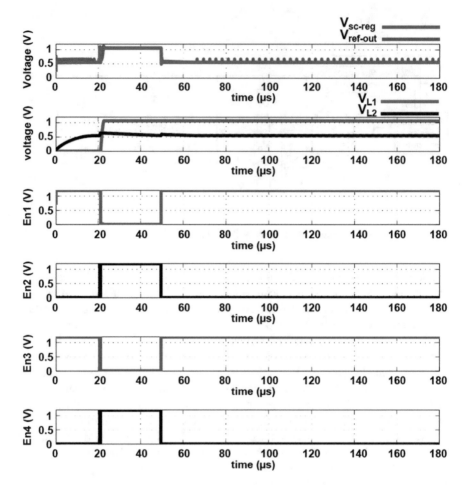

Fig. 4.4 Simulation of DOSC at no load condition

waveforms in addition to the switches' configuration table for each gain setting. Figure 4.5c shows the SC configuration when the gain is 1, where C_{fly1-2} are charged by the input voltage V_{in} during ϕ_1. During ϕ_2, C_{fly1-2} discharges their energy into the output capacitor of C_{sc-reg}. On the other hand, Fig. 4.5d illustrates the SC configuration when the gain is 1/2. During ϕ_1, C_{fly1-2} are charged to $V_{in} - V_{sc-reg}$. During ϕ_2, C_{fly1-2} are discharged into the C_{sc-reg} resulting in an output voltage of $V_{sc-reg} = 1/2V_{in}$. There are three main metrics that affect the design of the SC regulator: power efficiency, voltage ripple, and area. To improve the ripple, higher flying and load capacitors are required, but this results in a larger area. Increasing the frequency would also improve the voltage ripple. However, the high clock frequency can reduce power efficiency if it is constantly fixed and not modulated according to the load current. The design choice of the flying capacitors,

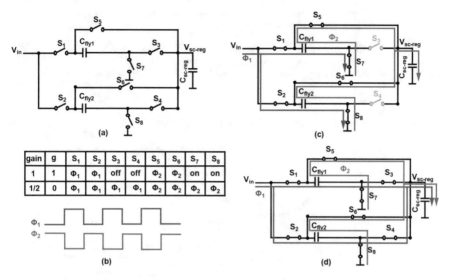

Fig. 4.5 (**a**) SC circuit, (**b**) SC switches configuration table including two phase signals waveform, (**c**) SC circuit at $g = 1$, (**c**) SC circuit at $g = 1/2$

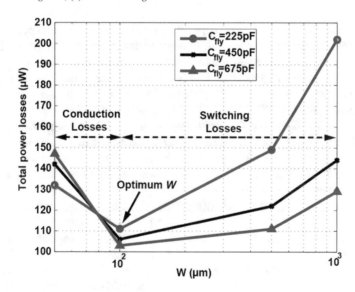

Fig. 4.6 Power losses of SC circuit versus W for different values of C_{fly}

the load capacitor, transistor width, and clock frequency has been discussed in the first author's master thesis [11] and a paper publication [10].

In order to design an efficient SC regulator, W, C_{fly}, and f_{sw} should be optimized so that the power losses are minimized. Figure 4.6 shows the total power losses of the SC circuit versus the switch widths W for various flying capacitors.

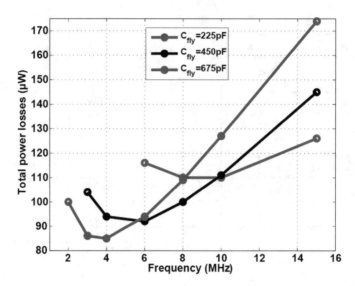

Fig. 4.7 Power losses of SC circuit versus f_{sw} for different values of C_{fly} at $W = 100\,\mu\text{m}$

The switch width is swept from $50\,\mu\text{m}$ till 1 mm to calculate the total power losses. As depicted in Fig. 4.6, the optimum switch width is achieved at $W = 100\,\mu\text{m}$ where the total losses record the minimum of $100\text{--}110\,\mu\text{W}$ depending on C_{fly}. This is because the switching losses are highly affected by the switch size according to Eq. (2.11) so that small transistor width results in low switching losses. However, the total losses increase at $W = 50\,\mu\text{m}$ because conduction losses become dominant. In order to select C_{fly}, Fig. 4.7 shows the effect of the switching frequency upon the total power losses for three different C_{fly} values at a maximum load current. As depicted in the figure, although the minimum power loss when $C_{fly} = 675\,\text{pF}$ is $1.3\times$ less than when $C_{fly} = 225\,\text{pF}$, the power density of the latter is $3\times$ higher than the former.

Therefore, the smallest capacitance of 225 pF is chosen. In addition, the choice of capacitance used for both flying and load capacitors has been carefully analyzed to reduce the bottom plate parasitic losses. Even though MIM capacitor reduces the bottom plate capacitor losses, it has a small capacitance density of $0.92\,\text{fF}/\mu\text{m}^2$. However, MOS capacitor provides higher capacitance density of $2.45\,\text{fF}/\mu\text{m}^2$ but adds more bottom plate parasitic losses. Therefore, MIM capacitor has been selected to implement the total flying capacitors of 450 pF because they participate in the charging and discharging every clock cycle, and if the MOS capacitor is used, it will have a negative impact on the efficiency. The load capacitor of 1.67 nF utilizes MOS capacitance to achieve higher capacitance density. Moreover, since MIM capacitor only uses upper layer metals, the active area and lower level metal have been used by MOS capacitor. In order to select the appropriate value of f_{clk}, Fig. 4.8 shows the load voltage level while sweeping the clock frequency. As depicted in the figure, the load voltage levels of 1 V and 0.6 V are regulated at a clock frequency of 10 MHz,

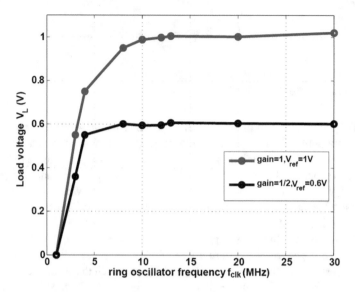

Fig. 4.8 Average load voltage levels versus the ring oscillator frequency

but due to PVT variations, 13 MHz has been selected. The proposed design adjusts the switching frequency depending on load current levels by the PFM. At heavy load current, the switching frequency is increased, whereas at light load current, the switching frequency is reduced since the switching losses dominate the total losses.

Changing the gain settings g of the SC regulator from 0 to 1 and vice versa, directly changes the intermediate node voltage of V_{sc-reg} from 0.5 to 1 V and vice versa. The main goal of changing the gain is to achieve a high power efficiency at its corresponding load voltage. The ideal power efficiency of the SC is given by Eq. (2.12). The ideal efficiency of the SC regulator at $V_{in} = 1.2$ V and $V_L = 0.55$ V when the gain is 1 is 45% compared to 83% when the gain is 1/2. Therefore, a 25% improvement in the power efficiency is achieved when varying the gain. However, changing the gain, which in turn changes V_{sc-reg} causes power losses across the flying capacitor. This loss is demonstrated in Eq. (4.1), where C represents the flying capacitors C_{fly1-2} and the output capacitor C_{sc-reg}, ΔV is the voltage swing across the capacitors, and $f_{sw-gain}$ is the switching frequency of DOSC between the two gains of 1 and 1/2 depending on the load current. Note that the DOSC runs at a frequency of 13 MHz but it does not switch between the gains every clock cycle. Based on the simulation results, the worst case switching frequency between the gains is 140 KHz which forms 1.1% of the DOSC frequency (13 MHz). However, the DOSC usually switches between the gains at a much lower frequency depending on the load current. Therefore, P_{loss} is compensated by the power efficiency improvement due to the gain change. Although the losses due to changing gain is reduced by a small amount of 1.2× compared with keeping the gain of the SC fixed, providing multiple gains is needed since our system operates most of the time at low voltage where the gain of 1/2 is required to support it efficiently.

$$P_{loss} = \frac{1}{2} C \Delta V^2 f_{sw\text{-}gain}$$ (4.1)

4.2.2   ATM Implementation

The ATM technique is implemented through FSM using ASIC design flow. Figure 4.9 shows the FSM diagram of the proposed ATM to eliminate the reverse current problem when switching from V_{L2} to V_{L1}. The FSM has three main states, namely S_0 (generates $V_{L2} = 0.55$ V), S_1 (prepares $V_{sc\text{-}reg} = 1$ V), and S_2 (generates $V_{L1} = 1$ V) with three different inputs of a, b, and c, where a is the output comparison between V_{L1} and V_{ref1}, b is the output comparison between V_{L2} and V_{ref2}, and c is the output comparison between $V_{sc\text{-}reg}$ and $V_{ref\text{-}out}$.

First, state S_0 has the highest priority to generate $V_{L2} = 0.55$ V. When $V_{L2} < V_{ref2}$, the FSM generates the digital bits of $En2 = En4 = g = 0$ and $En1 = En3 = 1$ (see Fig. 4.10 at x_1). Once V_{L2} and $V_{sc\text{-}reg}$ reach the steady state of 0.55 V while V_{L1} is below 1 V ($abc = 010$ at x_2 as in Fig. 4.10), the DOSC switches to S_1. During S_1, the FSM changes the digital configuration to $En1 = En2 = En4 = g = 1$ and $En3 = 0$. Changing to S_1 makes $V_{ref\text{-}out} = 1$ V whereas $V_{sc\text{-}reg}$ is still 0.55 V ($c = 1$ at x_3). Hence, the PFM increases $V_{sc\text{-}reg}$ from 0.55 to 1 V making $V_{sc\text{-}reg}$ almost equal to $V_{ref\text{-}out}$ ($c = 0$ at x_4) to eliminate the reverse current while both $M1$ and $M2$ are off. The off-time of $M1$ depends on how fast the PFM can bring $V_{sc\text{-}reg}$ to 1 V, which is controlled by the switching frequency of the regulator. The FSM switches to S_2 as long as V_{L2} remains at 0.55 V ($abc = 010$). This occurs when $En1 = En3 = 0$ and $En2 = En4 = g = 1$. Thus, the current flows forward since $V_{sc\text{-}reg} > V_{L1}$ until 1 V is generated at V_{L1}. In addition, if

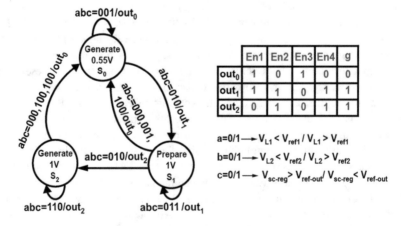

Fig. 4.9 FSM diagram of the ATM controller including three states: S_0 to generate 0,55 V, S_1 to raise $V_{sc\text{-}reg}$ from 0.55 to 1 V to eliminate reverse current and S_2 to generate 1 V

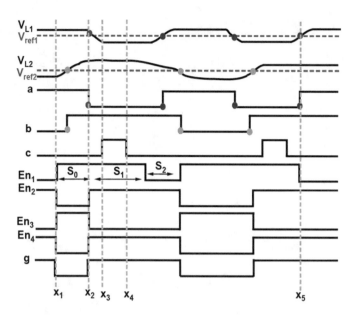

Fig. 4.10 Timing diagram of DOSC outputs and digital signals

both $V_{L1} = 1$ V and $V_{L2} = 0.55$ V, the DOSC remains in the last state; i.e., S_2 (see Fig. 4.10 at x_5); so that the switching activity and the losses are minimized. In contrast, if V_{L2} drops below 0.55 V during S_1, the DOSC returns to S_0 to regulate V_{L2} to 0.55 V. Refer to Appendix for the FSM Verilog code. The advantage of an ATM implementation is that it is a fully digital block that consumes low power and occupies a small area. The measurement results indicate that the power consumption of the ATM including the buffers to drive $M1$, $M2$, $M3$, $M4$, and g is 800 nW at maximum load condition (10 μA at 0.55 V and 250 μA at 1 V). Note that the FSM runs at a switching frequency of 13 MHz for the DOSC circuit.

Figure 4.11 shows the switching activity of DOSC between 0.55 and 1 V. If $V_{L2} = 0.55$ V and V_{L1} drops below 1 V as in Fig. 4.11a, then, the SC switches from 0.55 to 1 V. In order to eliminate the reverse current, both $M1$ and $M2$ are simultaneously deactivated as shown in Fig. 4.11b so that the PFM adjusts the SC voltage to increase V_{sc-reg} to 1 V. This way, the voltage droop due to the reverse current across V_{L1} vanishes (see Fig. 4.12). Note that the voltage droop is 80 mV in the presence of the reverse current as depicted in Fig. 4.12. This voltage droop degrades the performance if the reverse current problem is not addressed. In addition, switching frequency, load capacitor, and load current are the factors that affect the voltage droop. The response of the control technique should be fast enough when switching between the two voltage domains (1 V and 0.55 V) to reduce the voltage droop. In addition, if the size of the load capacitor is small and the load current is high, the voltage droop increases. Once V_{sc-reg} is stabilized and reaches 1 V, $M1$ is activated so that the current flows forward as shown in Fig. 4.11c. On the

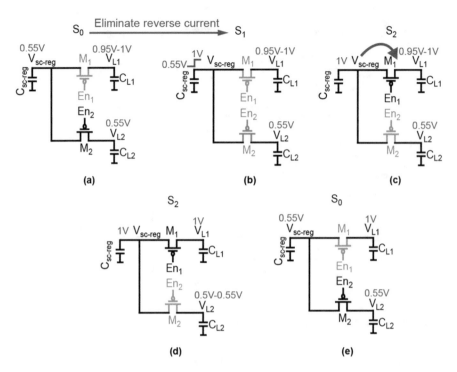

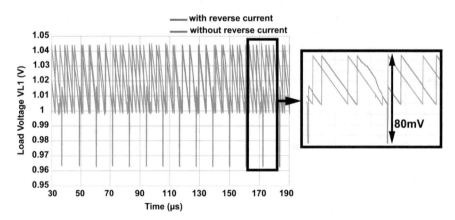

Fig. 4.11 (**a**) $V_{L2} = 0.55$ and $V_{L1} < 1$ V, (**b**) prepare $V_{sc\text{-}reg} = 1$ V, (**c**) forward current when switching from 0.55 to 1 V, (**d**) $V_{L1} = 1$ V and $V_{L2} < 0.55$ V, (**e**) forward current when switching from 1 to 0.55 V

Fig. 4.12 Simulated load voltage of V_{L1} with and without reverse current at the load of 5 μA at 0.55 V and 50 μA

other hand, Fig. 4.11d shows that if $V_{L1} = 1$ V and V_{L2} drops below 0.55 V, the SC switches from 1 to 0.55 V. As such, $M1$ is deactivated and $M2$ is activated where the charges redistribute between $C_{sc\text{-}reg}$ and C_{L2} as shown in Fig. 4.11e. Note that the current flow, in this case, is forward since $V_{L2} < V_{sc\text{-}reg}$.

4.2.3 Clocked Comparator

Figure 4.13 shows the clocked comparator circuit which is mainly based on latch amplifier. The operation of the clocked comparator depends on the amount of current passed through M_9 and M_{10}. As shown in Fig. 4.13, nodes x and y are precharged with the precharge drivers M_1, M_4, M_5, and M_8 when clk is low. Once clk gets high, the precharge drivers are turned off. If $V_{ref} > V_L$, then the drain source current $I_{ds,M_9} > I_{ds,M_{10}}$. This makes I_{ds,M_9} discharges faster into M_{11} than $I_{ds,M_{10}}$ decreasing node x to 0 V and implying $c = V_{in}$ and $c_b = 0$.

4.2.4 Ring Oscillator

The clock frequency depends on the inverter's time delay t_d as given in Eq. (4.2), where N is the odd number of inverters. Note that the transistor width is minimized and the transistor length is maximized for each inverter such that W/L = 150 nm/3μm. This is to reduce the power consumption of the ring oscillator as

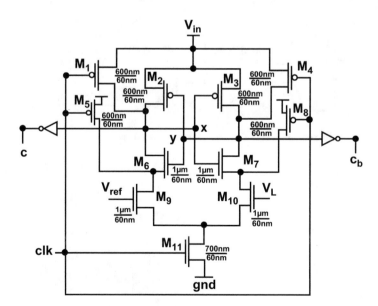

Fig. 4.13 Clocked comparator

it reaches $1.3\,\mu W$ at $V_{in} = 1.2\,V$. In addition, buffers are added to increase the strength of the ring oscillator.

$$f_{clk} = \frac{1}{2N \cdot t_d}. \tag{4.2}$$

4.2.5   Deadtime Phase Generator Circuit

A non-overlapping two-phase circuit that is known as a deadtime circuit is shown in Fig. 4.14. The deadtime circuit creates a time t_{dead} between ϕ_1 and ϕ_2 where all switches of the SC regulator are off for a short period of time. This is important to avoid any short circuit current from V_{dd} to ground during the switching.

4.3   Measured Results of DOSC Regulator in 65 nm CMOS Technology

The proposed DOSC regulator is fabricated in 65 nm CMOS. Figure 4.15 shows the layout of the DOSC where it occupies an area of $(0.418 \times 0.648)\,mm^2$. The total value of the MIM flying capacitors is 445 pF. Both MIM and MOS capacitances are utilized for the load capacitors to further reduce the DOSC area where $C_{sc\text{-}reg} = 100\,pF$ (MOS), $C_{L1} = 225\,pF$ (MIM) $+ 825\,pF$ (MOS), $C_{L2} = 225\,pF$ (MIM)

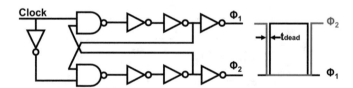

Fig. 4.14 Deadtime phase generator to eliminate short circuit current

Fig. 4.15 Layout of DOSC in 65 nm CMOS

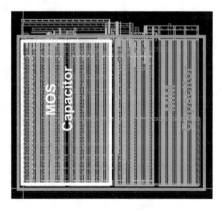

+ 275 pF (MOS). The size of the output capacitor $C_{sc\text{-}reg}$ is selected to support the required voltage without affecting the speed of the control loops. The DOSC regulator has an input voltage range from 1.05 to 1.4 V and operates at a clock frequency of 13 MHz. Figure 4.16 shows the die photo of the DOSC regulator with an area of 0.27 mm². The MOS capacitors are placed on the bottom of the MIM capacitors to reduce the area.

Figures 4.17 and 4.18 show the measured waveforms of two simultaneous regulated steady state output voltages of 1 V and 0.55 V at a minimum load condition ($I_{L2} = 1\,\mu\text{A}$ at $V_{L2} = 0.55\,\text{V}$ and $I_{L1} = 10\,\mu\text{A}$ at $V_{L1} = 1\,\text{V}$) and a

Fig. 4.16 Die photo of DOSC in 65 nm CMOS

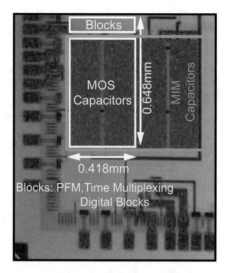

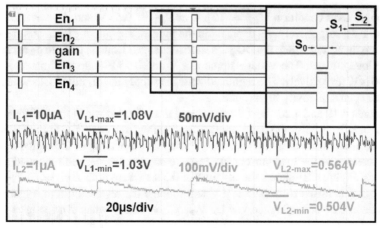

Fig. 4.17 Measured time domain waveforms of output voltages V_{L1} and V_{L2} and digital signals from FSM of $En1$, $En2$, $En3$, $En4$ and gain at minimum load of $I_{L1} = 10\,\mu\text{A}$ at $V_{L1} = 1\,\text{V}$ and $I_{L2} = 1\,\mu\text{A}$ at $V_{L2} = 0.55\,\text{V}$

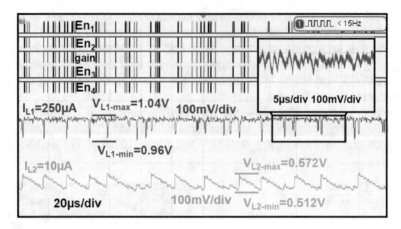

Fig. 4.18 Measured time domain waveforms of output voltages V_{L1} and V_{L2} and digital signals from FSM of $En1$, $En2$, $En3$, $En4$ and gain at maximum load of $I_{L1} = 250\,\mu\text{A}$ at $V_{L1} = 1\,\text{V}$ and $I_{L2} = 10\,\mu\text{A}$ at $V_{L2} = 0.55\,\text{V}$

Fig. 4.19 Measured digital signals from FSM of $En1$, $En2$, $En3$, $En4$ and modulated switching frequency of f_{sw} at minimum load of $I_{L1} = 10\,\mu\text{A}$ at $V_{L1} = 1\,\text{V}$ and $I_{L2} = 1\,\mu\text{A}$ at $V_{L2} = 0.55\,\text{V}$

maximum load condition ($I_{L2} = 10\,\mu\text{A}$ at $V_{L2} = 0.55\,\text{V}$ and $I_{L1} = 250\,\mu\text{A}$ at $V_{L1} = 1\,\text{V}$), respectively. As depicted in the figures, the output digital bits from the FSM are measured where the DOSC's state is changed from S_0 to S_1 to S_2 according to the load current. The voltage ripple at 1 V and 0.55 V is approximately 80 mV and 50 mV, respectively. The regulated voltages supply the digital blocks which can tolerate such a relativity high ripple.

Figures 4.19 and 4.20 show the measured time domain waveforms of the output digital bits from FSM and the modulated switching frequency f_{sw} at the minimum and maximum load current, respectively. Note that the activity of the digital bits and f_{sw} increases as the load current increases to support the desired regulated output voltages. Figure 4.21 shows the off-time of 0.5 μs where both $En1 = En2 = 1$ so that V_{sc-reg} is raised from 0.55 to 1 V to eliminate the reverse current. Figure 4.22 indicates the output reference voltage $V_{ref-out} = 0.55\,\text{V}$ when $En4 = g = 0$ and $En3 = 1$ and; $V_{ref-out} = 1\,\text{V}$ when $En4 = g = 1$ and $En3 = 0$.

Figure 4.23 shows the measured transient response of DOSC at three different conditions of load current. The first condition as shown in Fig. 4.23a is when I_2 is

Fig. 4.20 Measured digital signals from FSM of *En*1, *En*2, *En*3, *En*4 and modulated switching frequency of f_{sw} at maximum load of $I_{L1} = 250\,\mu A$ at $V_{L1} = 1\,V$ and $I_{L2} = 10\,\mu A$ at $V_{L2} = 0.55\,V$

Fig. 4.21 Measured time domain waveforms when both M_1 and M_2 are off for $0.5\,\mu s$ to raise up $V_{sc\text{-}reg}$ to $1\,V$ before turning on M_1

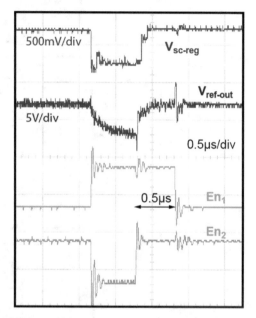

Fig. 4.22 Measured time domain waveforms when $V_{L2} = 0.55\,V$, En_3 is logic high (M_3 is off), En_4 is logic low (M_4 is on), and gain is logic low (voltage gain = 1/2) and when $V_{L1} = 1\,V$, En_3 is logic low (M_3 is on), En_4 is logic high (M_4 is off), and gain is logic high (voltage gain = 1)

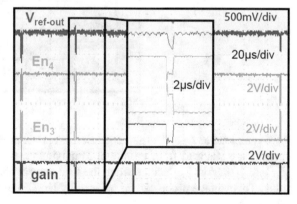

changed from the minimum value of $1\,\mu A$ to the maximum value of $10\,\mu A$ where there is no load at I_{L1}. The second condition as shown in Fig. 4.23b is when I_{L1} is changed from the minimum value of $10\,\mu A$ to the maximum value of $250\,\mu A$ where

Fig. 4.23 Measured transient response at three different conditions of load current (**a**) I_{L2} changes from 1 to 10 μA and $I_L = 0$, (**b**) I_{L1} changes from 10 to 250 μA and $I_{L2} = 0$, (**c**) I_{L2} changes from 1 to 10 μA and I_{L1} changes from 10 to 250 μA

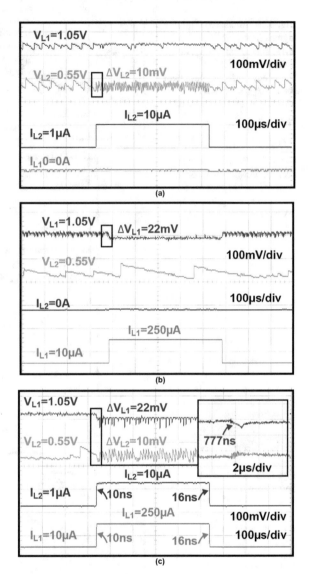

there is no load at I_{L2}. The third condition as shown in Fig. 4.23c is when both load currents are changed from minimum to maximum. A voltage droop of 22 mV in V_{L1} and 10 mV in V_{L2} occurs when the load current changes from a low to a high value. This voltage droop can be reduced by increasing the frequency and the load capacitors.

Figure 4.24 shows the measured power efficiency versus the load current at three different input voltage levels: 1.05 V, 1.2 V, and 1.4 V. The peak efficiency of 78%

Fig. 4.24 Measured power
efficiency versus load current
at (**a**) $V_{in} = 1.05$ V, (**b**)
$V_{in} = 1.2$ V, and (**c**)
$V_{in} = 1.4$ V

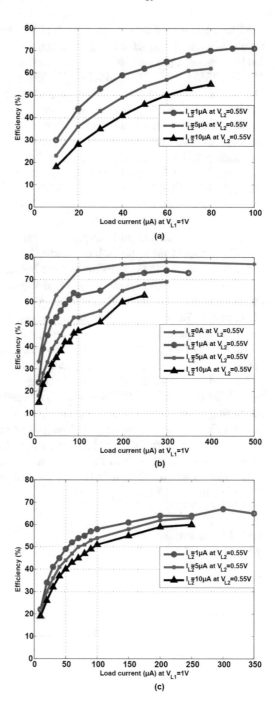

is achieved at a load current of 300 μA at 1 V and no load current at 0.55 V and nominal input voltage of 1.2 V. In addition, 74% is achieved at $V_{in} = 1.2$ V and load currents of 1 μA and 300 μA at load voltages of 0.55 V and 1 V, respectively. At $V_{in} = 1.05$ V, the maximum efficiency is 70% at load currents of 1 μA and 100 μA at load voltages of 0.55 V and 1 V, respectively. When $V_{in} = 1.05$ V, the load current at 1 V decreases to 100 μA. The maximum power efficiency at $V_{in} = 1.4$ V is 67% at load currents of 1 μA and 100 μA at load voltages of 0.55 V and 1 V, respectively. Note that the efficiency among all three graphs increases as I_{L2} decreases since the switching activity of the DOSC decreases. The efficiency of the DOSC is similar to the efficiency of the LDO regulator at the gain of 1, whereas at the gain of 1/2, the efficiency of the DOSC is improved and the efficiency enhancement factor (EEF) is calculated as 0.25 ($EEF = 1 - \frac{\eta LDO}{\eta SC}$ [12]). However, at a light load current of 10 μA, the efficiency of the DOSC decreases due to the fixed power losses from the control circuits. One option to improve the efficiency of the PMU is to bypass the DOSC and supply the required blocks from the energy source.

Table 4.1 shows the calculations of the breakdown power losses in the DOSC regulator at $V_{L1} = 1$ V, $I_{L1} = 250$ μA and $V_{L2} = 0.55$ V, $I_{L2} = 5$ μA. The overall digital control circuits that include digital gain controller, deadtime generator, ATM, PFM, ring oscillator have 16% of power losses. P_{Clfy} which includes the losses at gain equals 1, gain equals 1/2, and the change between the two gains has the largest power losses of 46% whereas P_{cond} forms 30% of the overall power losses. Finally, the power loss due to switch gate capacitance P_{sw} is 8.1%.

Table 4.2 shows the comparison between state of the art and the proposed work. The proposed DOSC design shows an excellent area efficiency improvement, supporting a μW power range that is suitable for wearable biomedical devices. The power efficiency in [8] in the μW load power range is less than 50% which is lower compared to our DOSC. Moreover, the power density of the proposed DOSC regulator is much better than the design in [2] that targets a similar load power range. The proposed DOSC is fully implemented on-chip unlike the work in [3] and [7] which utilizes off-chip capacitors.

Table 4.1 Break down of the losses in DOSC at $V_{L1} = 1$ V, $I_{L1} = 250$ μA and $V_{L2} = 0.55$ V, $I_{L2} = 5$ μA

Source of losses	Value
$P_{digital}$	5.4%
P_{ATM}	2.6%
P_{PFM}	4.1%
P_{M1+M2}	4.1%
P_{Cfly}	46%
P_{sw}	8.1%
P_{cond}	30%

Table 4.2 Comparison between prior and proposed works

Source	[2]	[3]	[8]	[7]	This work
Process	180nm CMOS	45nm CMOS	28nm CMOS	65nm CMOS	65nm CMOS
Topology	Parallel SC	Ladder SC	Dynamic power cells Allocation	Parallel SC with time multiplexing	DOSC
Vin	0.9V–4V	1V	1.3V–1.6V	0.85V–3.6V	1.05V–1.4V
VL	0.6V,1.2V, 3.3V	0.66V,0.33V	0.4V–0.9V	0.1V–1.9V	0.55V,1V
ηmax	81% @PL=15.9μW,Vin=2.8V	90%[a] @ PL=1.3mW, Vin=1V	83.3% @PL=45.9mW,Vin=1.5V	95.8% @VL1=VL2=1.2V,Vin=2.5V IL=10mA	78% @ PL=300μW, Vin=1.2V
Cfly	3nF (MIM+MOS)	3.7nF (on-chip)	8.1nF(on-chip) (MOM+MOS)	6nF (on-chip) @SC1 1μF (off-chip) @SC2	450pF (MIM)
CL	not reported	1.5nF (off-chip)	not reported	10μF at each output	1.6nF (MIM+MOS)
Load	20nW-500μW @1.2V, 3.3V, 0.6V	500μA-5mA @0.66V 10μA-1mA @0.33V	0-100mA@VL1 100mA-0@VL2	50μA-10mA @1V and 1.2V	10μA-350μA @1V 1μA-10μA @0.55V
Area	2.36mm²	0.37mm²	0.6mm²	3mm²	0.27mm²
Power density	0.211mW/mm²	0.0099mW/mm²	150mW/mm²	7.33mW/mm²	1.32mW/mm²

[a]Simulation

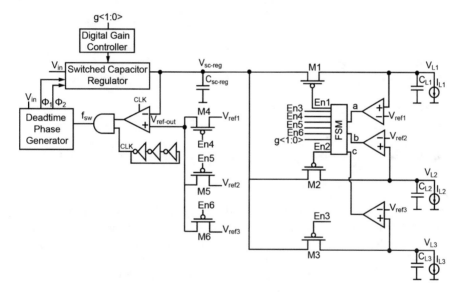

Fig. 4.25 Block diagram of multioutput SC (MOSC) regulator supporting three outputs

4.4 Multi-Outputs Switched Capacitor Voltage Regulator

The previous design of the DOSC can be modified to generate more than two outputs. Figure 4.25 shows the multi-outputs SC (MOSC) regulator that provides three output voltage levels of $V_{L1} = 1\,V$, $V_{L2} = 0.7\,V$, and $V_{L3} = 0.55\,V$. The ATM is also modified where an additional comparator is added and the FSM is programmed in a way to generate three outputs. The FSM generates multiple digital signals. $En1$ and $En4$ control $V_{L1} = 1\,V$, $En2$ and $En5$ control $V_{L2} = 0.7\,V$, and

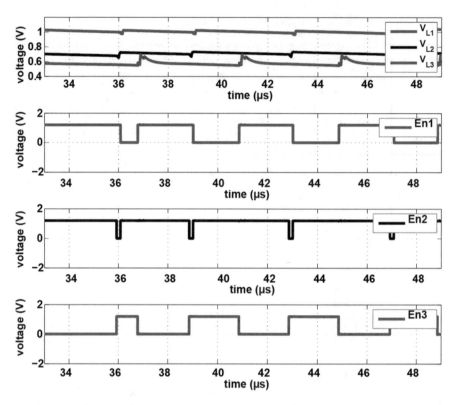

Fig. 4.26 Simulation results of MOSC that generates three outputs $V_{L1} = 1$ V, $V_{L2} = 0.7$ V, and $V_{L3} = 0.55$ V controlled by $En1$, $En2$, and $En3$, respectively

$En3$ and $En6$ control $V_{L3} = 0.55$ V. In addition, $g < 1 : 0 >$ controls the voltage gain of the SC regulator depending on the desired V_L value. For example, when V_{L1} is required, then, the gain $= 1$ ($g < 1 : 0 >= 11$) whereas when V_{L2} is required, the gain $= 2/3$ ($g < 1 : 0 >= 01$). Moreover, when V_{L3} is required, the gain is changed to $1/2$ ($g < 1 : 0 >= 00$). Figure 4.26 shows the simulation result of the MOSC where the three output voltages are generated and controlled by their corresponding signals. Comparing the MOSC that generates three output voltages with the DOSC that generates two output voltages, the area of the MOSC is increased $1.39\times$ since an additional load capacitor is added. In addition, the power efficiency of the MOSC is reduced to 30% at a load power of 140 μW compared to power efficiency of 50% for the DOSC. This means that the power efficiency of the MOSC is reduced by $1.66\times$ which also depends on the load power. This is because the switching between the outputs increases which increases the switching losses.

4.5 Summary

A dual-outputs switched capacitor voltage regulator in 65 nm CMOS has been presented. The proposed design has two main blocks: the switched capacitor regulator and the finite state machine. The switched capacitor regulator generates an output voltage of 1 V or 0.55 V from an input voltage range between 1.05 and 1.4 V, based on a reference voltage. The finite state machine utilizes an adaptive time multiplexing technique to simultaneously generate two output voltages of 1 V and 0.55 V to support a load current range between 10–350 μA and 1–10 μA, respectively. The finite state machine is designed to eliminate the reverse current that is created when switching occurs between the two output voltages. The measured results indicate that the peak efficiency of 78% is achieved at 300 μW (300 μA at 1 V and no load at 0.55 V) with an area of 0.27 mm^2. At a light load current, the dual-outputs switched capacitor can be bypassed to improve the efficiency of the PMU. The dual-outputs switched capacitor has been designed to generate two output voltages and it can be modified to provide more than two outputs at cost of area and power efficiency.

Appendix: Verilog Code of ATM

```
module FSM (En1,En2,En3,En4,g,a,b,c,clk,rst,state,next-state); input a, b, c ;
input rst, clk;
output reg En1, En2, En3, En4, g;
output reg [1:0] state;
output reg [1:0] next-state;
parameter [1:0] S0=2'b00, S1=2'b01, S2=2'b11;
always @ (a or b or c or state)
begin
case (state)
S0: case (a,b,c)
3'b000: begin next-state=S0; En1= 1; En2= 0; g=0; En3=1; En4=0; end
    3'b001: begin next-state=S0; En1= 1; En2= 0; g=0; En3=1; En4=0; end
    3'b010: begin next-state=S1; En1= 1; En2= 1; g=1; En3=0; En4=1; end
    3'b100: begin next-state=S0; En1= 1; En2= 0; g=0; En3=1; En4=0; end
    3'b101: begin next-state=S0; En1= 1; En2= 0; g=0; En3=1; En4=0; end
    3'b110: begin next-state=S0; En1= 1; En2= 0; g=0; En3=1; En4=0; end
    default: begin next-state=S0; En1= 1; En2= 0; g=0; En3=1; En4=0; end
    endcase
    S1: case (a,b,c)
3'b000: begin next-state=S0; En1= 1; En2= 0; g=0; En3=1; En4=0; end
    3'b001: begin next-state=S0; En1= 1; En2= 0; g=0; En3=1; En4=0; end
    3'b010: begin next-state=S2; En1= 0; En2= 1; g=1; En3=0; En4=1; end
```

```
    3'b011: begin next-state=S1; En1= 1; En2= 1; g=1; En3=0; En4=1; end
    3'b100: begin next-state=S0; En1= 1; En2= 0; g=0; En3=1; En4=0; end
    3'b110: begin next-state=S1; En1= 1; En2= 1; g=1; En3=0; En4=1; end
    default: begin next-state=S1; En1= 1; En2= 1; g=1; En3=0; En4=1; end
    endcase
    S2: case (a,b,c)
3'b000: begin next-state=S0; En1= 1; En2= 0; g=0; En3=1; En4=0; end
    3'b001: begin next-state=S0; En1= 1; En2= 0; g=0; En3=1; En4=0; end
    3'b010: begin next-state=S2; En1= 0; En2= 1; g=1; En3=0; En4=1; end
    3'b011: begin next-state=S2; En1= 0; En2= 1; g=1; En3=0; En4=1; end
    3'b100: begin next-state=S0; En1= 1; En2= 0; g=0; En3=1; En4=0; end
    3'b110: begin next-state=S2; En1= 0; En2= 1; g=1; En3=0; En4=1; end
    3'b110: begin next-state=S2; En1= 0; En2= 1; g=1; En3=0; En4=1; end
    default: begin next-state=S2; En1= 0; En2= 1; g=1; En3=0; En4=1; end
    endcase
    endcase end
    always @ (posedge clk or negedge rst)
    if (rst==0)
state<=S0;
    else state<=next-state;
```

References

1. Z. Hua and H. Lee, "A reconfigurable dual-output switched-capacitor DC-DC regulator with sub-harmonic adaptive-on-time control for low-power applications," *IEEE Journal of Solid-State Circuits*, vol. 50, no. 3, pp. 724–736, 2015.
2. W. Jung, J. Gu, P. D. Myers, M. Shim, S. Jeong, K. Yang, M. Choi, Z. Foo, S. Bang, S. Oh *et al.*, "A 60%-efficiency 20 nw-500 μw tri-output fully integrated power management unit with environmental adaptation and load-proportional biasing for IoT systems," in *Solid-State Circuits Conference (ISSCC), 2016 IEEE International*. IEEE, 2016, pp. 154–155.
3. Y. Zhao, Y. Yang, K. Mazumdar, X. Guo, and M. R. Stan, "A multi-output on-chip switched-capacitor DC-DC converter for near-and sub-threshold power modes," in *Circuits and Systems (ISCAS), 2014 IEEE International Symposium on*. IEEE, 2014, pp. 1632–1635.
4. S. Ahsanuzzaman, A. Prodić, and D. A. Johns, "An integrated high-density power management solution for portable applications based on a multioutput switched-capacitor circuit," *IEEE Transactions on Power Electronics*, vol. 31, no. 6, pp. 4305–4323, 2016.
5. P. Kumar and W. Proefrock, "Novel switched capacitor based triple output fixed ratio converter (TOFRC)," in *Applied Power Electronics Conference and Exposition (APEC), 2012 Twenty-Seventh Annual IEEE*. IEEE, 2012, pp. 2352–2356.
6. Z. Safarian and H. Hashemi, "Capacitance-sharing, dual-output, compact, switched-capacitor DC–DC converter for low-power biomedical implants," *Electronics Letters*, vol. 50, no. 23, pp. 1673–1675, 2014.
7. C. K. Teh and A. Suzuki, "A 2-output step-up/step-down switched-capacitor DC-DC converter with 95.8% peak efficiency and 0.85-to-3.6 V input voltage range," in *Solid-State Circuits Conference (ISSCC), 2016 IEEE International*. IEEE, 2016, pp. 222–223.
8. J. Jiang, Y. Lu, W.-H. Ki, U. Seng-Pan, and R. P. Martins, "A dual-symmetrical-output switched-capacitor converter with dynamic power cells and minimized cross regulation for

application processors in 28 nm CMOS," in *Solid-State Circuits Conference (ISSCC), 2017 IEEE International*. IEEE, 2017, pp. 344–345.

9. T. Instruments. (2016) Reverse Current Protection in Load Switches. [Online]. Available: http://www.ti.com/lit/an/slva730/slva730.pdf

10. D. Kilani, M. Alhawari, B. Mohammad, H. Saleh, and M. Ismail, "An Efficient Switched-Capacitor DC-DC Buck Converter for Self-Powered Wearable Electronics," *IEEE Transactions on Circuits and Systems I: Regular Papers*, vol. 63, no. 10, pp. 1557–1566, 2016.

11. D. Kilani, "An efficient on-chip switched-capacitor DC-DC converter for ultra-low power applications."

12. T. Van Breussegem and M. Steyaert, *CMOS integrated capacitive DC-DC converters*. Springer Science & Business Media, 2012.

Chapter 5
Ratioed Logic Comparator-Based Digital LDO Regulator

The proposed wearable biomedical device as given in Fig. 1.12 includes analog and noise sensitive blocks that should be supplied by LDO regulator in order to support a clean and a low voltage ripple. LDO regulators are widely utilized in PMU as they provide fast response, small area, and full integration. However, with the increase of process variation and the reduction of the supply voltage, the analog amplifier becomes challenging to design. As such, many research focus moves towards digital LDO (DLDO) regulator design as it can operate at a low supply voltage level. Yet digital logic circuits introduce more delay which affects the transient response. This chapter discusses the state of the art DLDO regulators in the literature. Then, we propose a DLDO regulator based on a ratioed logic comparator circuit that totally eliminates the digital loop delay. After that, we present the simulation results in 22 nm FDSOI technology.

5.1 State of the Art Digital LDO Regulator

Different DLDO regulator designs are discussed in the literature based on the way that the power switch is controlled. Figure 5.1a shows the architecture of the synchronous shift register (SR)-based control DLDO design. This circuit is considered as the conventional DLDO design that was first published in [1]. As shown in the figure, the synchronous DLDO consists of a clocked comparator, an n-bit SR, and an n-power switch. The number of the power switches n depends on the required load current I_L. If the load voltage V_L is lower than the reference voltage V_{ref}, the output of the comparator will be logic low and the output of the synchronous SR Q[n:0] are shifted toward right in order to increase the number of the turned-on power switches. Once the PMOS power switches are turned on, the PMOS current I_M starts increasing. In contrast, if V_L is higher than V_{ref}, the output

© Springer Nature Switzerland AG 2020
D. Kilani et al., *Power Management for Wearable Electronic Devices*, Analog Circuits and Signal Processing, https://doi.org/10.1007/978-3-030-37884-4_5

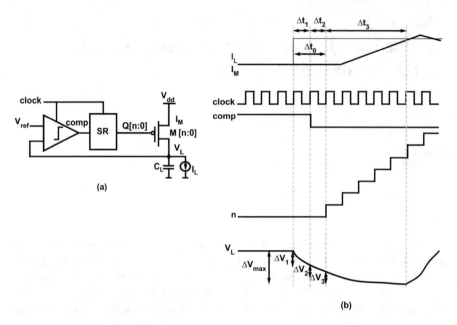

Fig. 5.1 (a) Synchronous SR DLDO, (b) waveform of the conventional synchronous SR DLDO in the undershoot condition

of the comparator will become logic high and all bits are shifted toward the left to decrease the number of the turned-on power switches.

The synchronous SR shifts one bit per clock cycle that consequently turns on/off one power switch per clock cycle. This affects the DLDO speed and response during the load transient as shown in Fig. 5.1b where two time delays are introduced. The first time delay Δt_1 is the time that the comparator takes to react when I_L steps up with a short transition. The second time delay Δt_2 is the time that the synchronous SR takes to react to turn on the power switch. Hence, the synchronous DLDO requires a total of two clock cycles Δt_o to respond to the load transient. This time delay causes a voltage droop at the output. The maximum voltage droop ΔV_{droop} is determined in Eq. (5.1) where ΔV_1 is the voltage droop during Δt_1, ΔV_2 is the voltage droop during Δt_2, and ΔV_3 is the voltage droop during the time Δt_3 that SR takes to turn on the target number of the switches. In addition, ΔV_{droop} is not only a function of time but also a function of I_L and load capacitor C_L as given in Eq. (5.2). One way to minimize the voltage droop is to increase C_L at the expense of area. Another way is to increase the clock frequency, however, the power increases. Therefore, the demand to improve the transient speed response while maintaining high power efficiency and small area is desired.

$$\Delta V_{droop} = \Delta V_1 + \Delta V_2 + \Delta V_3 \tag{5.1}$$

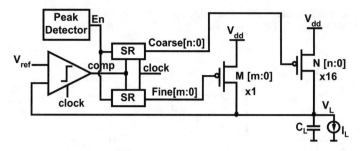

Fig. 5.2 Coarse-fine tuning (CFT) DLDO where high switching frequency and large transistor widths are utilized in coarse loop and low switching frequency and small transistor widths are utilized in fine loop [2]

$$\Delta V_{droop} = \frac{I_L \Delta t_o}{C_L} \tag{5.2}$$

To solve the speed-power tradeoff, coarse-fine tuning (CFT) technique is used where larger power switches can be turned on/off during the transient load [2]. Figure 5.2 shows the CFT-DLDO design that consists of two main loops: coarse and fine. The coarse loop is enabled during the transient load to turn on/off larger transistor width ×16 and higher clock frequency of 500 MHz. On the other hand, the fine loop is enabled with smaller transistors' width ×1 and lower clock frequency of 50 MHz during the steady state condition. The CFT technique enhances transient speed by reducing Δt_{1-3}. However, this reduction still depends on increasing the frequency which increases the power.

To reduce Δt_1, the work in [3] eliminates the use of the clocked comparator and introduces a logic-threshold triggered comparator (LLTC) that triggers the difference between the reference voltage and output voltage based on the inverter characteristics. The inverter switches between logic 1 and 0 based on the logic-threshold voltage V_{lth} that is determined by the aspect ratio of the PMOS and NMOS transistor as given in Eq. (5.3) where μ_n and μ_p are the NMOS and PMOS transistor mobility, respectively, W_n and W_p are the NMOS and PMOS transistor width, respectively, and V_{Tn} and V_{Tp} are the NMOS and PMOS threshold voltage, respectively. Note that the value of V_{lth} linearly scaled with the supply voltage V_{dd}.

$$V_{lth} = \frac{V_{dd}\sqrt{\frac{\mu_p W_p}{\mu_n W_n}} + V_{Tn} - \sqrt{\frac{\mu_p W_p}{\mu_n W_n}}|V_{Tp}|}{1 + \sqrt{\frac{\mu_p W_p}{\mu_n W_n}}} \tag{5.3}$$

Figure 5.3a shows the LLTC circuit design. It consists of a diode-connected inverter that is supplied by V_{ref} and followed by a series of two inverters which are supplied by V_L. The diode-connected inverter ($M1, M2$) works as a voltage divider and generates a fixed voltage $V_{ref-div}$ from V_{ref} which is fed to the gate of the first inverter ($M3, M4$). The first inverter is sized in a way to toggle when

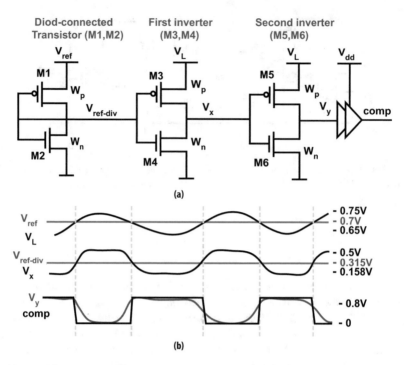

Fig. 5.3 (a) Clockless comparator using LLTC circuit; (b) waveform of previous work of LLTC [3]

$V_{lth} = V_{ref\text{-}div}$. The value of V_{lth} changes according to the value of supply voltage which is V_L. Figure 5.3b shows the waveform of the LLTC where $V_{ref} = 0.7$ V, $V_{ref\text{-}div} = 0.315$ V. Whenever V_L crosses V_{ref}, the first inverter toggles the output V_x high (when $V_L > V_{ref}$) or low (when $V_L < V_{ref}$). However, both $M3$ and $M4$ are not fully on/off which disallows V_x to completely reach full V_L or gnd. As shown in Fig. 5.3b, the output of the first inverter V_x fluctuates between 0.5 and 0.158 V. Therefore, a second inverter is required so that V_y can fully reach V_L or gnd. When $V_L > V_{ref}$, $M5$ is off and $M6$ is in the linear region whereas when $V_L < V_{ref}$, $M5$ is linear and $M6$ is off. This way ensures that the output of the second inverter V_y swings between full V_L and gnd. The disadvantage of this design is that the V_{gs4} is always fixed which affects the response of the first inverter as it only depends on V_{gs3}. In addition, in the beginning, $V_L = 0$ which disallow the first and second inverters to operate and fail to generate a *comp* signal. Therefore, a startup circuit is required.

On the other hand, to reduce Δt_2, the synchronous SR is replaced by the asynchronous control unit as shown in Fig. 5.4 [4, 5]. The work in [4] is the first design that introduces the idea of the asynchronous DLDO. However, asynchronous logic circuits are more sensitive to PVT variations as the delay increases significantly at subthreshold supply voltage [6]. The asynchronous DLDO work in [5] has been

Fig. 5.4 Asynchronous
DLDO

Fig. 5.5 Hybrid LDO using
high pass filter analog loop
and SR digital loop [10]

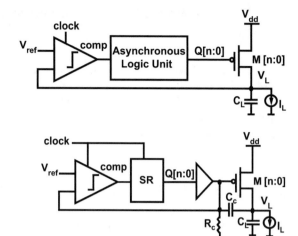

designed for low supply voltage operating point including CFT technique. The asynchronous logic units are used in the coarse loop and the barrel SR is used in the fine loop. Although this work reduces ΔV_2 (by using asynchronous logic unit) and ΔV_3 (by using CFT technique), ΔV_1 still exists due to the usage of the clocked comparator.

Alternatively, ADCs were utilized in [7, 8] to implement multi-bit quantization technique so that multiple power switches are turned on/off per one clock cycle. The work in [7] utilizes 7-bit inverter-based ADC and event-driven control technique rather than time driven technique to minimize the load transient time response. The work in [8] implements a flash ADC that removes the delay control loop of the DLDO. In this case, the speed of the DLDO is determined by the speed of the flash ADC. However, the speed-power tradeoff still exits since the design suffers from the large quiescent current. Further, voltage-to-time converter (VTC) has been used in [9] by converting V_{ref} and V_{out} into pulses ϕ_{ref} and ϕ_{out}, respectively, where each pulse duty corresponds to the voltage amplitude. Then, the phase detector measures the time difference between ϕ_{ref} and ϕ_{out} that corresponds to a voltage difference between V_{ref} and V_{out}. This time difference is digitized through the time-to-digital converter (TDC) and then stored into UP/DN counter that turns on/off the required number of power switches. Based on the digitized time difference, the counter turns on/off a certain number of power switches. Recent works implement hybrid LDO regulator where both analog and digital control techniques are utilized in [10–12]. The design in [10] implements an additional high pass analog loop in parallel with a digital loop in order to achieve a faster response as shown in Fig. 5.5. The advantage of this design is that the analog loop responds continuously to the load transient before the digital loop. The design in [12] replaces the PMOS power switch with the NMOS since it provides higher current during the load transient when the load voltage drops. However, it requires a charge pump that adds extra power budget to overdrive the NMOS transistor.

5.2 Proposed Ratioed Logic Comparator-Based Digital LDO Regulator

In this section, we will first discuss our proposed ratioed logic comparator (RLC)-based DLDO (RLC-DLDO). The proposed RLC is an enhancement circuit of the LTTC design in [3]. It has less number of inverters, lower power consumption, and smaller area. Then, we will explain the limitation of the RLC-DLDO design and discuss possible improvements.

5.2.1 Ratioed Logic Comparator-Based DLDO Circuit Design

As explained in the previous section, the clocked comparator is widely used in DLDO to reduce the power consumption but at the expense of the regulator's transient response. As an alternative solution, we propose RLC-DLDO as shown in Fig. 5.6a. It consists of an asynchronous digital RLC circuit and a PMOS power switch M_{p0}. The RLC circuit generates a signal *comp* to turn on/off M_{p0} based on the comparison between V_{ref} and V_L. It totally eliminates the usage of the clock and it reacts continuously to any changes in V_L. Therefore, one power switch is enough to provide V_L unlike an array of PMOS power switches in the conventional design. This is due to the small range of load current that we are targeting which is between 10 and 500 μA.

Figure 5.6b shows the waveform of the proposed RLC-DLDO during the transient load. When I_L steps up with a short transition, the RLC circuit senses the change at V_L and generates continuously $comp = 0$ to turn on M_{p0}. Consequently, the PMOS transistor current I_{p0} starts increasing to support the required load current and raise V_L to its required voltage level. The speed of the RLC-DLDO is determined by the speed of the RLC circuit. This means that the maximum time

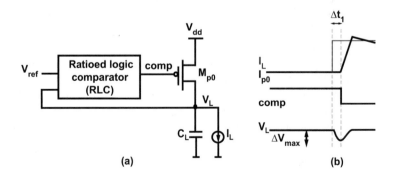

Fig. 5.6 (a) Proposed RLC-DLDO using clockless RLC circuit; (b) waveform of proposed RLC-DLDO during transient load

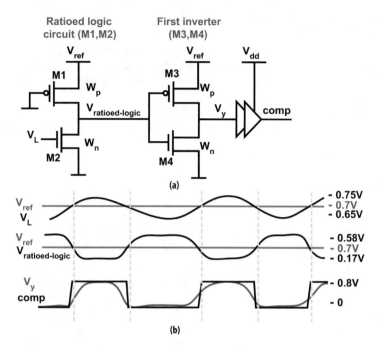

Fig. 5.7 (**a**) RLC circuit; (**b**) waveform of RLC circuit

that the RLC-DLDO consumes is $\Delta t_0 = \Delta t_1$ where Δt_1 is time taken by the RLC to respond to the load transient as shown in Fig. 5.6b. Hence, ΔV_{droop} is reduced to ΔV_1 compared to ΔV_{droop} in Eq. (5.1) in the conventional design. This leads to a faster response and lower undershoot/overshoot output voltage.

Figure 5.7a shows the proposed RLC circuit. It consists of a ratioed logic circuit ($M1$, $M2$) and an inverter ($M3$, $M4$) which are both supplied by V_{ref}. The ratioed logic circuit includes a PMOS transistor $M1$ and a pseudo NMOS transistor $M2$. $M1$ is always on where the gate terminal is connected to gnd. The gate of $M2$ is connected to V_L that controls the output voltage of the ratioed logic circuit $V_{ratioed\text{-}logic}$. Initially, when $V_L = 0$, $M2$ is off and $M1$ is in the linear region so that $V_{ratioed\text{-}logic}$ is pulled up making $V_y = comp = 0$. When V_L starts increasing and becomes above V_{ref}, the source-drain resistance of M2 R_{ds2} decreases which decreases $V_{ratioed\text{-}logic}$. The low output voltage level of $V_{ratioed\text{-}logic}$ is given in Eq. (5.4) and it strongly depends on V_L, V_{ref} and the aspect ratio of $M1$ and $M2$ ($\frac{\beta_p}{\beta_n} = \frac{\mu_p C_{ox} \frac{W_p}{L_p}}{\mu_n C_{ox} \frac{W_n}{L_n}}$). Conversely, when V_L drops below V_{ref}, R_{ds2} increases and hence $M1$ pulls up $V_{ratioed\text{-}logic}$. The high output voltage level of $V_{ratioed\text{-}logic\text{-}high}$ is given in Eq. (5.5) and it also depends on V_L, V_{ref}, and $\frac{\beta_n}{\beta_p}$ ratio. Refer to Appendix for the derivation of $V_{ratioed\text{-}logic\text{-}low}$ and $V_{ratioed\text{-}logic\text{-}high}$.

Fig. 5.8 Transfer characteristics of the (**a**) ratioed logic circuit ($M1$, $M2$); (**b**) inverter ($M3$, $M4$)

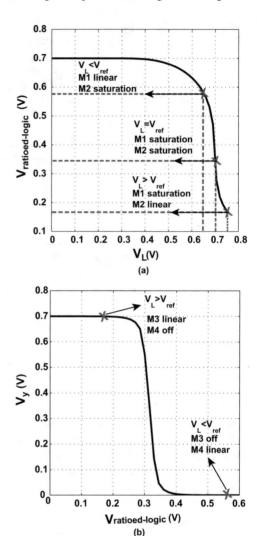

$$V_{ratioed\text{-}logic\text{-}low} = (V_L - V_{Tn}) - \sqrt{(V_L - V_{Tn})^2 - \frac{\beta_p}{\beta_n}(V_{ref} - V_{Tp})^2} \qquad (5.4)$$

$$V_{ratioed\text{-}logic\text{-}high} = V_{Tp} + \sqrt{(V_{ref} - V_{Tp})^2 - \frac{\beta_n}{\beta_p}(V_L - V_{Tn})^2} \qquad (5.5)$$

To better understand the functionality of the RLC, Fig. 5.7b shows the proposed design waveform where $V_{ref} = 0.7$ V, V_L is between 0.75 and 0.65 V. When V_L increases above 0.7 V, let us say 0.75 V, $M2$ gets stronger and becomes in linear whereas $M1$ is in saturation as shown in Fig. 5.8a. This makes $V_{ratioed\text{-}logic}$ decrease to 0.17 V. On the other hand, when V_L decreases below 0.7 V, let us say 0.65 V, $M1$ gets stronger and becomes in linear whereas $M2$ in saturation as shown in Fig. 5.8a. This increases $V_{ratioed\text{-}logic}$ to 0.58 V. When $V_L = V_{ref}$, both $M1$ and $M2$ are in saturation. In order to achieve such operation, $M1$ and $M2$ should be sized carefully. Note that $V_{ratioed\text{-}logic}$ does not reach V_{ref} and gnd since both $M1$ and $M2$ operate either in saturation or linear regions. Thus, an inverter is required to pull up/down the output V_y to V_{ref} or gnd. A buffer is added in order to drive the output of RLC *comp*. Figure 5.8b depicts the transfer characteristics of the inverter and it shows that when $V_L > V_{ref}$, $M3$ is in linear region whereas $M4$ is off so that $V_y = V_{ref} = 0.7$ V. In addition, when $V_L < V_{ref}$, $M3$ is off whereas $M4$ is in linear region which implies $V_y = 0$.

Comparing our proposed RLC design with the LLTC design given in Fig. 5.3 [3], RLC has less number of transistors of 4 compared with 6 in LLTC. Hence the power consumption of the RLC decreases. Table 5.1 shows the comparison between the RLC and the LLTC when simulating both of them in 22 nm FDSOI at $V_{dd} = 0.8$ V and $V_L = 0.7$ V. The transistor's size of both circuits is given in Table 5.2. Note that the sizing of the LLTC is taken from [3]. As shown in Table 5.1, the current consumption of the RLC is reduced by 2× compared with the LLTC. This is because one inverter is utilized in the RLC unlike the LLTC where two inverters are implemented which increases the DC current and switching losses. Further, RLC does not have a startup issue when $V_L = 0$ since it is fed to the gate of $M2$ unlike in LLTC where V_L is the main source of the circuit.

In addition, RLC requires less sizing effort as only 2 transistors ($M1$, $M2$) require sizing compared to 6 transistors ($M1 - M6$) in LLTC. Since the gate voltage of the first inverter ($M3$, $M4$) in the RLC is variable, both V_{gs3} and V_{gs4} changes according to V_L unlike the LLTC where only V_{gs3} changes. This has more impact on the speed and the gain of the inverter. However, this advantage does not improve the speed compared to LLTC. In RLC, V_L is first fed back to the gate of the ratioed logic circuit and its output goes to the first inverter. On the other hand, in the LLTC, V_L is fed back directly to the source of the first inverter and its output goes to the second inverter. Therefore, both designs have two stages and almost have a similar response time.

Table 5.1 Comparison between RLC and LLTC circuits at $V_{dd} = 0.8$ V and $V_L = 0.7$ V

Parameter	LLTC	RLC
Transistors' number	6	4
Current consumption	2.8 μA	1.4 μA
Startup	Yes	No

Table 5.2 Transistors' size of the LLTC and RLC at $V_{dd} = 0.8$ V and $V_L = 0.7$ V

Transistor	LLTC	RLC
M_1	100 nm/100 nm	80 nm/2 μm
M_2	100 nm/100 nm	80 nm/2.5 μm
M_3, M_4	130 nm/20 nm	80 nm/20 nm
M_5, M_6	130 nm/20 nm	–

However, the RLC-DLDO given in Fig. 5.6 faces some challenges. First, the proposed design can support one V_L level from one V_{ref}. Thus, the design should be adaptive to support more than one load voltage level. Hence, the size of $M1$ and $M2$ in the ratioed logic circuit should be adaptively changed since $V_{ratioed\text{-}logic}$ depends on the aspect ratio of $M1$ and $M2$, V_L and V_{ref} as given in Eqs. (5.4) and (5.5). Therefore, the size of $M1$ and $M2$ varies according to the value of V_{ref}. Third, the ratioed logic circuit suffers from a large static current because there will be always a path from the supply to the ground. Therefore, the channel length of $M1$ and $M2$ should be large enough to reduce the static current.

5.2.2   Enhanced RLC-DLDO

Figure 5.9 shows the block diagram of the proposed RLC-DLDO regulator. It consists of RLC circuit, PMOS power switches M_{p0-3}, and control unit. The RLC generates a digital bit *comp* based on the comparison between V_{ref} and V_L. The *comp* signal is connected to the gate of M_{p0} to either turn it off or on depending on the comparison output. To support different V_L levels of 0.7 V, 0.6 V, and 0.5 V from V_{dd} of 0.8 V, 0.7 V, and 0.6 V, respectively, M_{p1-3} are utilized with different transistor widths to provide load current between 10 and 500 μA. These M_{p1-3} transistors are controlled by the control unit that generates $\sim d_{1-3}$ from x and y. When $V_L = 0.7$ V is required, M_{p3} is enabled by $\sim d_3 = 0$ whereas when $V_L = 0.6$ V is required, M_{p2} is enabled by $\sim d_2 = 0$. In addition, when $V_L = 0.5$ V is needed, M_{p1} is enabled by $\sim d_1 = 0$. The digital bit *en* ensures that M_{p1-3} are off at the beginning before the operation of the regulator. For example, when *en* = 0,

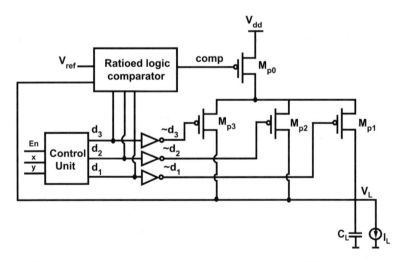

Fig. 5.9 Block diagram of the proposed ratioed logic comparator-based DLDO

Table 5.3 Input–output combinations of the control unit in the RLC-DLDO regulator

en	xy	$\sim d_1 \sim d_2 \sim d_3$
0	XX	111
1	00	011
1	10	101
1	11	110

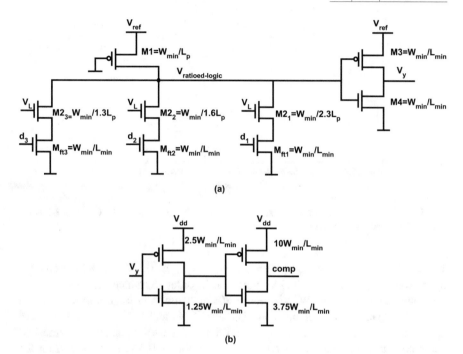

Fig. 5.10 (a) Proposed RLC circuit and its sizing; (b) buffer circuit and its sizing

the decoder control unit generates $\sim d_{1-3} = 1$ to make sure that M_{p1-3} are off. The digital signals x, y and en are provided from the SoC based on the required voltage domain. Table 5.3 shows the input–output combinations of the control unit.

The control unit also generates d_{1-3} to control the RLC output based on V_L or V_{ref}. Figure 5.10a shows the RLC circuit diagram. The ratioed logic circuit consists of a pull up network of PMOS transistor $M1$ and a pull down network of NMOS transistors $M2_{1-3}$ and M_{ft1-3}. $M1$ and $M2_{1-3}$ generates $V_{ratioed\text{-}logic}$ whereas M_{ft1-3} are three footer transistors controlled by d_{1-3} based on V_{ref}. As the proposed design supports three output voltage levels, three reference voltage levels are required. Since $V_{ratioed\text{-}logic}$ depends on the supply voltage V_{ref} as given in Eqs. (5.4) and (5.5), the aspect ratio of $M1$ and $M2_{1-3}$ has to be different for different V_{ref} value. Hence, $M2_{1-3}$ have different transistor size as shown in Fig. 5.10a. These transistors $M2_{1-3}$ are enabled by the footer transistors M_{ft1-3} which are controlled by d_{1-3}. For example, $M2_1$ is enabled by d_1 when $V_{ref} =$

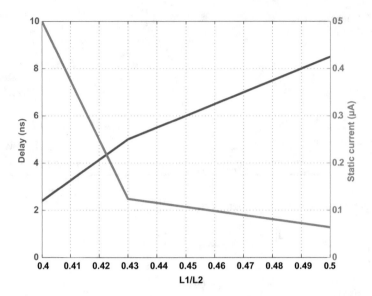

Fig. 5.11 Delay and static current versus $L1/L2_1$ at $V_L = 0.5V$ of the RLC circuit. $L1$ is the length of $M1$ and $L2_1$ is length of $M2_1$

$V_L = 0.5$ V, $M2_2$ is enabled by d_2 when $V_{ref} = V_L = 0.6$ V and $M2_3$ is enabled by d_3 when $V_{ref} = V_L = 0.7$ V. Note that in order to reduce the RLC current, minimum transistor size ($W_{min}/L_{min} = 80$ nm/20 nm) and large transistor length ($L_p = 1$ μm) have been implemented. The size of $M1$ and $M2$ is selected to achieve optimum static current and transient load delay. Figure 5.11 shows the transient delay and static current versus $L1/L2_1$ ratio at $V_L = 0.5$ V. As shown from the figure, the static current decreases as $L1/L2_1$ increases at the cost of the transient delay. The ratio of $L1/L2_1 = 0.43$ is selected at $V_L = 0.5$ V where the static current $= 0.124$ nA and the transient delay is 5 ns.

In order to ensure the functionality of the RLC circuit to trigger correctly at the required V_L, the size of $M1$ and $M2$ significantly plays an important role. Thus, the length of $M2_{1-3}$ is swept and optimized to generate the desired $V_{ratioed-logic}$. Figure 5.12 shows the transfer characteristics of the RLC circuit for different $M2_3$ channel length $L2_3$ values at $V_{ref} = 0.7$ V. Note that the channel length of $M1$ $L1$ is fixed at 1 μm. As shown in the figure, as $L2_3$ increases up to 2 μm, $M2_3$ becomes weaker and $V_{ratioed-logic}$ is still high at high V_L. This high $β_p/β_n$ ratio affects $V_{ratioed-logic-low}$ which prevents the RLC from switching to the low voltage level. At $L2_3 = 1$ μm, the load line switches before V_L reaches 0.7 V so that $V_{ratioed-logic}$ becomes low before $V_L = 0.7$ V. This means that the comparison is done before $V_L = 0.7$ V which might lead to an inaccurate result. At $L2_3 = 1.3$ μm, the load line switches at $V_L = 0.7$ V. Hence, $L2_3 = 1.3$ μm is selected at $V_{ref} = 0.7$ V. The same analysis is implemented at $V_{ref} = 0.6$ V and $V_{ref} = 0.5$ to properly select $L2_2$ and $L2_1$, respectively. Figure 5.10b shows the circuit diagram of a buffer that drives FO8.

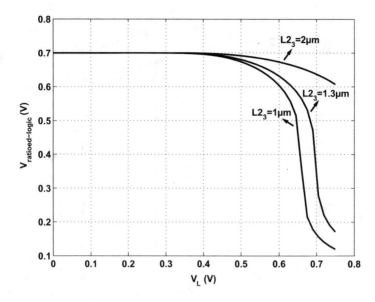

Fig. 5.12 Transfer characteristics of the RLC circuit at $V_{ref} = 0.7$ V for different $L2_3$ values at $L_1 = 1\,\mu$m

5.3 RLC-DLDO Design Optimization

In order to select the appropriate type of PMOS transistors M_{p0-3} that support the load current, Fig. 5.13 shows the IV characteristics of the super low threshold voltage PMOS transistor (slvt-pmos) and the standard PMOS transistor (rvt-pmos) that has a nominal threshold voltage. As shown in the figure, the slvt-pmos transistor has more capability to generate a current than the rvt-pmos transistor. For example, at $V_L = 0.7$ V, the slvt-pmos transistor supports higher current by 35% than the rvt-pmos transistor. Hence, the slvt-pmos provides more output current when V_L drops which further improve the speed of the transient response. Therefore, the slvt-pmos is selected for M_{p0-3} in the RLC-DLDO.

Each power switch M_{p1-3} has different transistor widths that support I_L between 10 and 500 μA for different load voltages. For each V_L level, an equivalent PMOS transistor width is selected. The equivalent width is the total transistor width of M_{p0} and M_{p1} or M_{p0} and M_{p2} or M_{p0} and M_{p3}. Figure 5.14 shows the load voltage versus the equivalent PMOS transistor width W_{eq} for different supply voltages at $I_L = 500\,\mu$A. As shown in the figure, at $V_{dd} = 0.8$ V, the required transistor width to support $V_L = 0.7$ V is $W_{eq} = 1.5\,\mu$m; at $V_{dd} = 0.7$ V, the required transistor width to support $V_L = 0.6$ V is $W_{eq} = 2\,\mu$m; and at $V_{dd} = 0.6$ V, the required transistor width to support $V_L = 0.5$ V is $W_{eq} = 3\,\mu$m. Figure 5.15 shows the output voltage ripple versus load current for different load capacitor C_L values at $V_{dd} = 0.8$ V and $V_L = 0.7$ V. $C_L = 100$ pF has been selected since it provides low output voltage ripple and occupies a small area.

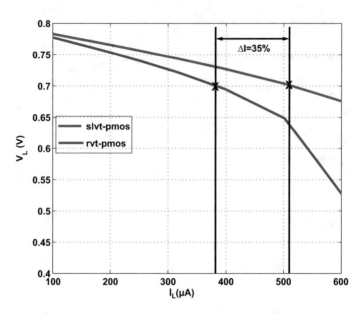

Fig. 5.13 IV characteristics of the turned-on super low threshold voltage (slvt) and standard transistor (rvt) PMOS at $V_{dd} = 0.8$ V and W/L $= 1.5\,\mu$m/20 nm

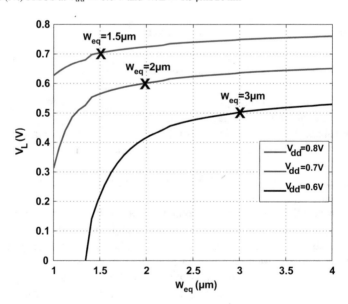

Fig. 5.14 PMOS transistor width for different V_{dd} at maximum load current $I_L = 500\,\mu$A

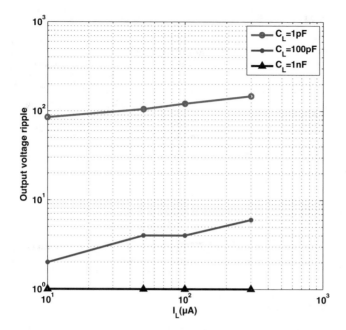

Fig. 5.15 Output voltage ripple versus load current for different load capacitor values at $V_{in} = 0.8$ V and $V_L = 0.7$ V

Fig. 5.16 Layout of the RLC-DLDO

5.4   Simulation Results of RLC-DLDO Regulator in 22 nm FDSOI

The RLC-DLDO regulator circuit is designed, implemented, and simulated in 22 nm FDSOI technology. Figure 5.16 shows the layout of the RLC- DLDO regulator. It occupies an area of $0.017\,\mathrm{mm^2}$. A MOM capacitor is used in the load capacitor to minimize the leakage current.

Figure 5.17 shows the load voltage–load current characteristics of the RLC-DLDO regulator. As shown from the figure, the load regulation is $20\,\mu V/\mu A$, $12\,\mu V/\mu A$, and $2\,\mu V/\mu A$ at $V_L = 0.7$ V, $V_L = 0.6$ V, and $V_L = 0.5$ V, respectively. Figure 5.18 shows the input–output voltage characteristics of the RLC-DLDO regulator. As shown from the figure, the line regulation is 2 mV/V and 3 mV/V at $V_{ref} = 0.6$ V and $V_{ref} = 0.5$ V, respectively.

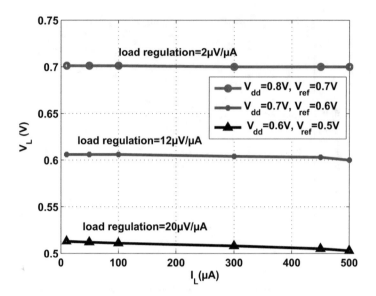

Fig. 5.17 V_L versus I_L of the RLC-DLDO

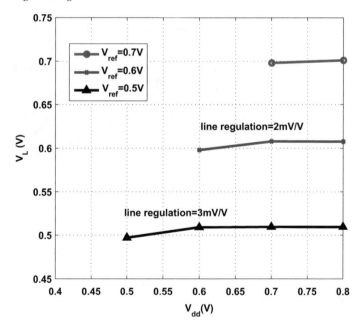

Fig. 5.18 V_L versus V_{dd} of the RLC-DLDO at $I_L = 10\,\mu A$

Figure 5.19 shows the transient load simulation between $I_L = 10\,\mu A$ and $I_L = 370\,\mu A$ at $V_L = 0.5\,V$. At high load current of $370\,\mu A$, the RLC circuit has a higher switching frequency of $45\,MHz$ than the lower load current of $10\,\mu A$

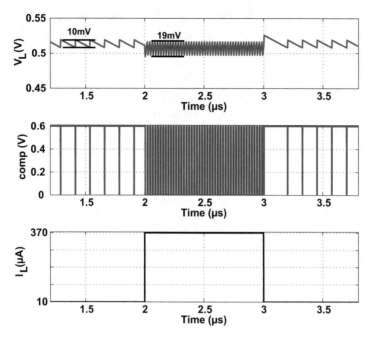

Fig. 5.19 Load transient simulation of RLC-DLDO between $I_L = 10\,\mu\text{A}$ and $I_L = 370\,\mu\text{A}$ at $V_{dd} = 0.6\,\text{V}$ and $V_L = 0.5\,\text{V}$

of 8 MHz. This enhances the current efficiency at light load current where the switching losses are reduced rather than having a fixed amount of power coming from the clock in the conventional design. Figures 5.20 and 5.21 show the overshoot and undershoot transient load condition, respectively, between 10 and 370 μA at $V_L = 0.5\,\text{V}$. As shown from the figure, the overshoot and undershoot voltages are 25 mV and 16 mV, respectively, when the load steps up or down between 10 and 370 μA. In addition, $\Delta t_1 = 5\,\text{ns}$ when the RLC responds to the step up/down load condition. Comparing this delay value with the conventional synchronous SR DLDO in [1] that has a clock frequency of 1 MHz and targets a similar load current in the μA, the transient response is reduced from μs to ns. In addition, the current consumption is improved by 9× since the 1 MHz clock generator consumes 2.7 μA compared to the overall current consumption (static current of ratioed logic circuit + switching current of the inverters and buffer) of 0.3 μA in the proposed RLC-DLDO at light load current of 10 μA. However, as the load current increases to 370 μA, the switching current of inverter and buffer increases to 0.8 μA but it is still better compared to the conventional DLDO in [1]. Figure 5.22 shows the current efficiency of the RLC-DLDO regulator. The maximum current efficiency is 99.9%.

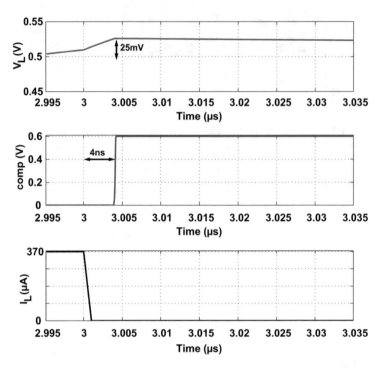

Fig. 5.20 Overshoot condition of the RLC-DLDO from $I_L = 370\,\mu$A and $I_L = 10\,\mu$A at $V_{dd} = 0.6$ V and $V_L = 0.5$ V

Table 5.4 shows the comparison between our work and the state of the art DLDO regulator. Comparing with the work in [1] that targets similar load current range in the μA, we achieve better undershoot voltage with fully integrated design whereas their design requires off-chip output capacitor. The work in [2] suffers from a high quiescent current of $82\,\mu$A unlike our design where the maximum quiescent current is $0.8\,\mu$A. The voltage undershoot in [3, 5] and [10] is over 20% of the required voltage level unlike our design where the undershoot voltage is over 3%. The proposed RLC-DLDO design achieves better FOM than the design in [1] and [2] and approximately similar FOM to the design in [3]. The proposed design has been implemented in 65 nm CMOS technology in addition to the 22 nm FDSOI. Simulation results show that the FOM in 22 nm is smaller than in 65 nm because this technology has faster switching and higher performance which results in a lower transient load delay by 4× compared in 65 nm CMOS. Comparing the proposed DLDO with the design in [1] where both have been implemented in 65 nm CMOS targeting similar load current, the proposed DLDO achieves better FOM and current consumption reduction by 8×.

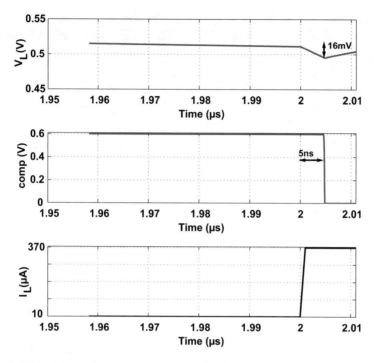

Fig. 5.21 Undershoot condition of the RLC-DLDO from $I_L = 10\,\mu\text{A}$ and $I_L = 370\,\mu\text{A}$ at $V_{dd} = 0.6\,\text{V}$ and $V_L = 0.5\,\text{V}$

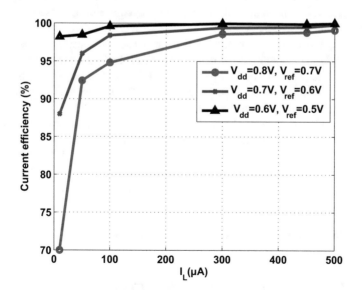

Fig. 5.22 Current efficiency of the RLC-DLDO

Table 5.4 Comparison table between the proposed DLDO and the state of the DLDO regulators

	[1]	[2]	[3]	[5]	[10]	This Work	
Technology	65nm CMOS	65nm CMOS	65nm CMOS	28nm CMOS	65nm CMOS	22nm FDSOI	65nm CMOS
Topology	Synchronous SR DLDO	CFT DLDO	Logic-threshold comparator	Asynchronous DLDO	Hybrid (analog+digial)	Ratioed Logic comparator	
V_{dd}	0.5V–1.2V	0.6V–1.1V	0.7V–1.2V	0.5V–1V	0.5V–1V	0.6V–0.8V	
V_L	0.45V	0.4V–1V	0.6V–1.1V	0.45V–0.95V	0.45V–0.95V	0.5V–0.7V	
C_L	0.1μF	1nF	1nF	0.1nF	0.1nF	0.1nF	
I_L	0-200μA	2mA–100mA	0A–23mA	2.5mA–33.5mA	2mA–12mA	10μA–500μA	
ΔV_L	undershoot=40mV overshoot=30mV	undershoot=55mV overshoot=70mV	undershoot=200mV	undershoot=101.7mV overshoot=70mV	undershoot=105mV overshoot=65mV	[a]undershoot=16mV overshoot=25mV	undershoot=49mV overshoot=36mV
Quiescent current	2.7μA	82μA	6μA	10μA	3.2μA	[b]min:0.195μA max: 0.4μA	[b]min:0.15μA max: 0.33μA
Line regulation	3.1mV/V	3mV/V	0.78mV/V	–	–	3mV/V	60mV/V
Load regulation	0.65mV/mA	0.065mV/mA	0.04mV/mA	–	–	min: 2μV/μA max: 20μV/μA	min: 27μV/μA
current efficiency	98.7%	99.92%	99.97%	99.97%	99.9%	99.9%	99.7%
Area	0.042mm²	0.01mm²	0.014mm²	0.008mm²	0.034mm²	0.017m²	–
FOM	0.27μs	7.75ps	2.17ps	0.11ps	0.23ps	2.27ps	3.94ps

[a]simulation   [b]Including static and switching currents   FOM=C_L*ΔV_L*I_q / I_L^2

5.5   Summary

This chapter has presented a fast and a current efficient ratioed logic comparator-based digital low drop out (RLC-DLDO) regulator that solves the speed-power tradeoff by eliminating the usage of the clock and the synchronous shift register (SR) in the conventional design. It achieves a transient speed improvement in the ns range and a quiescent current reduction by 9× over the conventional design that targets similar load current in the μA range. The RLC-DLDO consists of three main blocks: RLC, PMOS power switches, and control unit. The RLC compares between the reference and the load voltage and generates a single bit that turns on/off the PMOS switches. Unlike the clocked comparator, the RLC is an event-driven design that continuously responds to the voltage difference. The control unit provides digital bits to control the power switches and the RLC circuit in order to support different output voltage levels. The RLC-DLDO has an input voltage range between 0.8 and 0.6 V and generates an output voltage range between 0.7 and 0.5 V for load current between 10 and 500 μA. The design is implemented in 22 nm FDSOI technology and occupies an active area of 0.0171 mm². The simulation results show that the peak efficiency is 99.9% and the load transient response time is 5 ns at $V_L = 0.5$ V.

Appendix: Derivation of Output Voltage of the Ratioed Logic Circuit

Low Output Voltage of Ratioed Logic Circuit

Given the ratio logic circuit in Fig. 5.7, when $V_L > V_{ref}$, the PMOS transistor $M1$ is in the saturation region whereas the NMOS transistor operates in the linear region. This pulls down the output voltage of the ratioed logic circuit. The output voltage can be found by equating the linear drain–source current of NMOS transistor ($I_{ds,n\text{-}linear}$) with the saturation drain source current of PMOS transistor ($I_{ds,p\text{-}sat}$) as follows:

$$I_{ds,n\text{-}linear} = I_{ds,p\text{-}sat} \tag{5.6}$$

$$\beta_n[(V_{gsn} - V_{Tn})V_{dsn} - \frac{V_{dsn}^2}{2}] = \beta_p(V_{gsp} - V_{Tp})^2 \tag{5.7}$$

where $V_{gsn} = V_L$, $V_{dsn} = V_{ratioed\text{-}logic}$ and $V_{gsp} = V_{ref}$. This implies that:

$$[(V_L - V_{Tn})V_{ratioed\text{-}logic} - \frac{V_{ratioed\text{-}logic}^2}{2}] = \frac{\beta_p}{2\beta_n}(V_{gsp} - V_{Tp})^2 \tag{5.8}$$

$$\frac{V_{ratioed\text{-}logic}^2}{2} - (V_L - V_{Tn})V_{ratioed\text{-}logic} + \frac{\beta_p}{2\beta_n}(V_{gsp} - V_{Tp})^2 = 0 \tag{5.9}$$

Solving the quadratic equation of 5.9 where a=1/2, b=$-V_L+V_{Tn}$, c=$\frac{\beta_p}{2\beta_n}(V_{ref} - V_{Tp})^2$ implies:

$$V_{ratioed\text{-}logic} = (V_L - V_{Tn}) - \sqrt{(V_L - V_{Tn})^2 - \frac{\beta_p}{\beta_n}(V_{ref} - V_{Tp})^2} \tag{5.10}$$

High Output Voltage of Ratioed Logic Circuit

Given the ratio logic circuit in Fig. 5.7, When $V_L < V_{ref}$, the PMOS transistor $M1$ is in the linear region whereas the NMOS transistor operates in the saturation region. This pulls up the output voltage of the ratioed logic circuit. The output voltage can be found by equating the linear drain–source current of PMOS transistor ($I_{ds,p\text{-}linear}$) with the saturation drain source current of NMOS transistor ($I_{ds,p\text{-}sat}$) as follows:

$$I_{ds,p\text{-}linear} = I_{ds,n\text{-}sat} \tag{5.11}$$

$$\beta_p[(V_{gsp} - V_{Tp})V_{dsp} - \frac{V_{dsp}^2}{2}] = \beta_n(V_{gsn} - V_{Tn})^2 \tag{5.12}$$

where $V_{gsp} = V_{ref}$, $V_{dsp} = V_{ref} - V_{ratioed\text{-}logic}$ and $V_{gsn} = V_L$. This implies that:

$$[(V_{ref} - V_{Tp})(V_{ref} - V_{ratioed\text{-}logic}) - \frac{(V_{ref} - V_{ratioed-logic})^2}{2}] = \frac{\beta_n}{2\beta_p}(V_L - V_{Tn})^2$$

$$(5.13)$$

Let us say that: $V_1 = V_{ref} - V_{ratioed\text{-}logic}$ $V_2 = V_{ref} - V_{Tp}$
Substituting the above assumption in Eq. (5.13):

$$V_1 V_2 - \frac{V_1^2}{2} - \frac{\beta_n}{2\beta_p}(V_L - V_{Tn}^2) = 0 \qquad (5.14)$$

$$\frac{V_1^2}{2} - V_1 V_2 + \frac{\beta_n}{2\beta_p}(V_L - V_{Tn}^2) = 0 \qquad (5.15)$$

Solving the quadratic equation of 5.15 where $a = 1/2, b = -V_2$, and $c = \frac{\beta_n}{2\beta_p}(V_L - V_{Tn})^2$ implies:

$$V_{ratioed\text{-}logic} = V_{Tp} + \sqrt{(V_{ref} - V_{Tp})^2 - \frac{\beta_n}{\beta_p}(V_L - V_{Tn})^2} \qquad (5.16)$$

References

1. Y. Okuma, K. Ishida, Y. Ryu, X. Zhang, P.-H. Chen, K. Watanabe, M. Takamiya, and T. Sakurai, "0.5 V input digital LDO with 98.7% current efficiency and 2.7 μA quiescent current in 65 nm CMOS," in *Custom Integrated Circuits Conference (CICC), 2010 IEEE*. IEEE, 2010, pp. 1–4.
2. M. Huang, Y. Lu, S.-W. Sin, U. Seng-Pan, and R. P. Martins, "A fully integrated digital LDO with coarse–fine-tuning and burst-mode operation," *IEEE Transactions on Circuits and Systems II: Express Briefs*, vol. 63, no. 7, pp. 683–687, 2016.
3. M. A. Akram, W. Hong, and I.-C. Hwang, "Fast transient fully standard-cell-based all digital low-dropout regulator with 99.97% current efficiency," *IEEE Transactions on Power Electronics*, vol. 33, no. 9, pp. 8011–8019, 2018.
4. Y.-H. Lee, S.-Y. Peng, C.-C. Chiu, A. C.-H. Wu, K.-H. Chen, Y.-H. Lin, S.-W. Wang, T.-Y. Tsai, C.-C. Huang, and C.-C. Lee, "A low quiescent current asynchronous digital-LDO with PLL-modulated fast-DVS power management in 40 nm SoC for MIPS performance improvement," *IEEE Journal of Solid-State Circuits*, vol. 48, no. 4, pp. 1018–1030, 2013.
5. Y. Huang, Y. Lu, F. Maloberti, and R. P. Martins, "A dual-loop digital LDO regulator with asynchronous-flash binary coarse tuning," in *Circuits and Systems (ISCAS), 2018 IEEE International Symposium on*. IEEE, 2018, pp. 1–5.
6. Y. Li, X. Zhang, Z. Zhang, and Y. Lian, "A 0.45-to-1.2-v fully digital low-dropout voltage regulator with fast-transient controller for near/subthreshold circuits," *IEEE Transactions on Power Electronics*, vol. 31, no. 9, pp. 6341–6350, 2016.
7. D. Kim and M. Seok, "8.2 fully integrated low-drop-out regulator based on event-driven pi control," in *Solid-State Circuits Conference (ISSCC), 2016 IEEE International*. IEEE, 2016, pp. 148–149.

8. Y.-J. Lee, W. Qu, S. Singh, D.-Y. Kim, K.-H. Kim, S.-H. Kim, J.-J. Park, and G.-H. Cho, "A 200-ma digital low drop-out regulator with coarse-fine dual loop in mobile application processor," *IEEE Journal of Solid-State Circuits*, vol. 52, no. 1, pp. 64–76, 2017.

9. T.-J. Oh and I.-C. Hwang, "A 110-nm CMOS 0.7-V input transient-enhanced digital low-dropout regulator with 99.98% current efficiency at 80-mA load," *IEEE Transactions on Very Large Scale Integration (VLSI) Systems*, vol. 23, no. 7, pp. 1281–1286, 2015.

10. M. Huang, Y. Lu, U. Seng-Pan, and R. P. Martins, "An analog-assisted tri-loop digital low-dropout regulator," *IEEE Journal of Solid-State Circuits*, vol. 53, no. 1, pp. 20–34, 2018.

11. M. Huang, Y. Lu, and X. Lu, "Partial analogue-assisted digital low dropout regulator with transient body-drive and 2.5× FOM improvement," *Electronics Letters*, vol. 54, no. 5, pp. 282–283, 2018.

12. X. Ma, Y. Lu, R. P. Martins, and Q. Li, "A 0.4 v 430 nA quiescent current NMOS digital LDO with NAND-based analog-assisted loop in 28 nm CMOS," in *Solid-State Circuits Conference-(ISSCC), 2018 IEEE International*. IEEE, 2018, pp. 306–308.

Chapter 6
Conclusions and Future Work

In order to achieve batteryless operation of low power systems such as wearable devices, energy harvesting has been utilized to enable autonomous operation; however, it is critical to efficiently harvest energy from the surroundings. Furthermore, dynamic voltage scaling plays an important role in reducing the overall power consumption of the device. Both solutions require building a power management unit (PMU) that highly achieves energy-efficient systems with a minimum form factor to suit the wearable applications of the human body. Thus, this work has focused on designing efficient voltage regulators for TEG energy harvesting-based device.

6.1 Conclusions

Chapter 3 presents silicon characterizations of three various TEG-based PMU designs targeting μW electronic devices. The TEG-based PMU consists of two parts: power conversion and voltage regulation. In the power conversion, SI boost converter is utilized to extract the maximum power point and boost the TEG voltage from 50 to 65 mV into 0.6–1.4 V. In contrast, the voltage regulation block regulates the boosted voltage into two load voltages of 0.6 V and 0.8 V in order to achieve regulation and scaling over load currents of 10–100 μA. The three TEG-based PMU designs are: switched inductor boost converter followed by switched capacitor (SI+SC), inductor boost converter followed by LDO (SI+LDO), and inductor boost converter followed by switched capacitor and LDO (SI+SC+LDO). The peak efficiency of the three TEG-based PMUs varies between 60 and 65% at different TEG and load voltages. The voltage regulation range of the SC is affected by both the input voltage and the voltage gain. The SI+LDO design is a good option for low voltage ripple which is suitable for analog and RF blocks. The SI+SC design is preferable to support digital blocks since it can provide high voltage conversion

© Springer Nature Switzerland AG 2020

D. Kilani et al., *Power Management for Wearable Electronic Devices*, Analog Circuits and Signal Processing, https://doi.org/10.1007/978-3-030-37884-4_6

ratio. The SI+SC+LDO can support two output voltage levels from both SC and LDO regulators which can be used in both analog and digital blocks. SI+LDO and SI+SC+LDO designs have lower voltage ripple of 12 mV compared to 35 mV in SI+SC. The SI boost converter occupies an area of 0.036 mm^2 and it requires off-chip inductor. The fully integrated SC regulator occupies the largest area of 0.495 mm^2 whereas the LDO regulator occupies the smallest area of 0.0357 mm^2.

Chapter 4 presents area and power efficient dual-outputs switched capacitor (DOSC) DC–DC buck converter. The DOSC converter has an input voltage range between 1.05 and 1.4 V and generates two simultaneous regulated output voltages of 1 V and 0.55 V. The DOSC consists of two main blocks: a switched capacitor regulator and an adaptive time multiplexing (ATM) controller. The switched capacitor regulator generates a single regulated voltage using pulse frequency modulation based on a predetermined reference voltage. In addition, the ATM controller generates two simultaneous output voltages and eliminates the reverse current during the switching between the output voltages. Addressing the reverse current problem is important to reduce the output voltage droop and improve performance. The converter supports load currents of 10–350 μA and 1–10 μA at load voltages of 1 V and 0.55 V, respectively. The DOSC circuit is fabricated in 65 nm CMOS and it occupies an active area of 0.27 mm^2. Measured results show that a peak efficiency of 78% is achieved at a load power of 300 μW.

Chapter 5 presents fast and efficient digital LDO (DLDO) regulator utilizing a clockless ratioed logic comparator (RLC). In addition to eliminating the clock, the RLC-DLDO removes the shift registers used in the conventional DLDO. It achieves a transient speed improvement in the ns range and a current reduction by 9× over the conventional design that targets μA load current. The RLC-DLDO consists of RLC, PMOS power switches, and control unit. The RLC compares between the reference and the load voltage and generates a single bit that turns on/off the PMOS switches. Unlike the clocked comparator, the RLC is an event-driven design that continuously responds to the voltage difference. The control unit provides digital bits to control the power switches and the RLC circuit in order to support different output voltage levels. The RLC-DLDO has an input voltage range between 0.8 and 0.6 V and generates an output voltage range between 0.7 and 0.5 V for load current between 10 and 500 μA. The design is implemented in 22 nm FDSOI and occupies an active area of 0.0171 mm^2. The simulation results show that the peak efficiency is 99.9% and the load transient response time is 5 ns at $V_L = 0.5$ V.

6.2 Future Work

The amount of energy generated from the harvester might not be adequate to turn on the device. Therefore, monitoring the amount of the harvested energy is important to efficiently operate the wearable device at different power modes based on the available amount of energy from the harvester. Hence, an energy monitoring circuit and a control FSM are needed to be implemented. The FSM interfaces between energy monitoring and the SoC load. Based on the availability of the harvested energy, the FSM enables each block in the SoC to operate at the proper power mode. The characterization of the TEG-based PMU designs proposed in Chap. 3 consists of two or three cascaded stages: boost converter followed by voltage regulators. This affects the end-to-end power efficiency as it depends on the efficiency of each stage. Thus, the efficiency could be improved by implementing a single stage design that consists of a boost converter in addition to some control circuits to achieve voltage regulation. The switched capacitor buck converter design in Chap. 4 supports two simultaneous output voltage. The design can be further enhanced to support more output voltage levels depending on the required ones. In addition, the FSM in the ATM controller always generates two output voltages of 1 V and 0.55 V. However, in our system, the voltage level of 1 V is not required all the time and the voltage level of 0.55 V is always required. Therefore, the FSM needs to be enhanced to support dual or single output voltage depending on the power mode. The digital LDO regulator in Chap. 5 shows only simulation results. The design has been tapped out and fabricated in 22 nm FDSIO technology. The design needs to be characterized and measured to ensure the correct functionality.

Index

© Springer Nature Switzerland AG 2020
D. Kilani et al., *Power Management for Wearable Electronic Devices*, Analog Circuits
and Signal Processing, https://doi.org/10.1007/978-3-030-37884-4

Printed in the United States
By Bookmasters